IMAGE CONTENT RETARGETING

Maintaining Color, Tone, and Spatial Consistency

IMAGE CONTENT RETARGETING
Maintaining Color, Tone, and Spatial Consistency

Alessandro Artusi
Universitat de Girona, Spain

Tunç Ozan Aydın
Disney Research, Zürich, Switzerland

Olga Sorkine-Hornung
ETH Zürich, Switzerland

Francesco Banterle
ISTI-CNR, Pisa, Italy

Daniele Panozzo
ETH Zürich, Switzerland

CRC Press
Taylor & Francis Group
Boca Raton   London   New York

CRC Press is an imprint of the
Taylor & Francis Group, an **informa** business

AN A K PETERS BOOK

CRC Press
Taylor & Francis Group
6000 Broken Sound Parkway NW, Suite 300
Boca Raton, FL 33487-2742

Printed on acid-free paper
Version Date: 20160607

Printed and bound in India by Replika Press Pvt. Ltd.

International Standard Book Number-13: 978-1-4822-4991-0 (Hardback)

**Visit the Taylor & Francis Web site at
http://www.taylorandfrancis.com**

**and the CRC Press Web site at
http://www.crcpress.com**

[To all my family, always in my heart. - AA]

[To my nieces. - FB]

Contents

Foreword

When I was growing up in Spain, many more years ago than I'd like to admit, there was only one TV in my house. A black and white TV, with exactly two channels. They didn't really turn their brains into mush naming them: They were called Channel 1 and Channel 2. To be more precise, it was more like one and a half channels, since Channel 2 would only emit content from 7 PM until midnight. My favorite cartoons would be shown exactly once a week. And because recording devices were an outrageously expensive luxury item back then, if you missed them, they were gone. Forever. You would never, ever, be able to watch them again.

And boy, it was fantastic!

It didn't matter that I had to imagine the colors of Mazinger Z and its robot enemies from the gray-scale images on the TV. It also didn't matter that when an American movie was shown, either big chunks at the top and bottom of the screen would be pure black, or that the images would be badly distorted, in an effort to accommodate the different aspect ratios of the movie and the TV set. I didn't care about dynamic range, gamut mapping, color-to-gray conversions, all of which, of course, I had no idea they were even a thing. I had access to visual content, and that was cool!

Now times have changed. Images and movies are captured from many devices, and can be shown on many, many different displays. In color! We have large, 55-inch (and more!) screens; we have I-Max theaters; we have computer monitors, cellphones, tablets, projectors. We have canned and streamed video. We have a gazillion formats and compression algorithms to store all kinds of visual media. Color is not just color; we have many, many color appearance models. The combinatorial space of how my favorite childhood cartoons can be displayed is virtually infinite. But hold on a sec: How can we show the same content to different people with so many different display devices?

This is the process called *content retargeting*. And this book tells you how it is done. You will first learn about the key aspects of our visual perception that need to be taken into account when mapping visual media to different displays. Then color and dynamic range will be introduced; quoting my friend Erik Reinhard, currently with Technicolor, both can be considered a two sides of the same coin, so they are presented together for a proper, holistic view. How to turn a color image into a gray-scale image is the topic of the following chapter. Color is a higher-dimensional space, so some information needs to be discarded, and other information transformed. Now *how*, that is the issue. But things get even more interesting when you consider the role of style, and ask yourself the question of how to transfer it from one image to another. I have this great picture I took on holidays, but wouldn't it be great if I could somehow make it look a bit more like this other picture from this famous photographer? Maybe you have a clear mental depiction of the desired result, but unfortunately not all of us are that good at using commercial image manipulation software like Photoshop. What we want is to simply tell the computer, and let it produce the result automatically. Great, but how? Finally, the book deals with spatial retargeting, or how to accommodate visual media on the different sizes and aspect ratios of the many, many screens available to us. This has been a key research area in the past few years, with many works trying to puzzle this out. And finally, after everything is said and done, we are left with the question of OK, but how do we know if we have done a good job? How do we know if the resulting image, after all our transformations in many different spaces, is acceptable to a human observer? Enter the domain of image quality assessment, which will be thoroughly analyzed in the last chapter of the book.

All in all, this book is as timely as it gets, given the massive, massive amount of available visual media and the smorgasbord of display devices of very different characteristics that exist nowadays. It provides a cross section of what is possible today in image content retargeting, and a window to what may be possible tomorrow. If only I had known as a kid, I might have appreciated my Saturday morning Mazinger-Z shows even more.

Diego Gutierrez
September 2015
Zaragoza University

Preface

The availability on the market of a large variety of color display devices, characterized by significant differences in their color gamut, dynamic range, spatial resolution, etc., has brought a series of research challenges that have been a topic of discussion in the last 30 years. A typical problem is the acquisition of an image with a digital camera and its subsequent visualization on various media and display devices with different characteristics. If the acquired image is not processed in advance, the user experience will be to view the image on a display medium that may have large discrepancies in its colors and tone, dynamic range, and even image structure. Sometimes the color information of the image is discarded due to the constraints imposed by display devices with limited capability, i.e., gray-level media. The question is, therefore, how do we preprocess images to maximize the display quality on the particular device? Another classical problem is the need to transfer a specific set of colors, as well as the style, from one image to another one. The research problems and applications above can be classified as *retargeting* problems.

Retargeting is far from being an easy task. In many cases, it is even an ill-posed problem, such as when displaying standard dynamic range (SDR) images, typically 8 bits per color channel, on high dynamic range (HDR) displays. This problem requires the retargeting algorithm to generate new content which is absent in the original image. In this book, we present the latest solutions and techniques for retargeting images along various dimensions such as dynamic range, colors, and spatial resolution and aspect ratio. We offer, for the first time, a holistic view of the field. In addition, we discuss how to measure and analyze the changes applied to an image in terms of quality using both subjective methodologies, i.e., psychophysical experiments, and objective methodologies, i.e., computational metrics.

The book is divided into six chapters. Chapter 1 introduces the basic concepts that are necessary for further understanding of the topics treated in the subsequent chapters. These include details of the human eye mechanisms and the basic concepts of color science. Chapter 2 describes the overall *color rendering pipeline* for the traditional 8-bit depth per color channel images. This is followed by an in-depth discussion on how to extend the color rendering pipeline to color images that require a higher bit depth, such as *high dynamic range* images. These images typically have 32 bits per color channel in order to cover the vast luminance dynamic range and wide color gamut observed in the real world. To accommodate this new type of image in the color rendering pipeline, several issues need to be solved. Chapter 3 tackles the problem of converting a color image into a gray-scale image. After a brief introduction on typical dimensionality reduction problems, we review in detail the *color2gray* problem and existing approaches to solving it. Finally, the opposite problem, *colorization*, is introduced. It can be seen as part of the general problem of color mapping, but we prefer to clearly separate the *color transfer* from the colorization methods and present it as part of the *style transfer* problem in Chapter 4. The availability of displays with different spatial resolutions and aspect ratios has created the need to spatially retarget images in order to fit different displays. Chapter 5 investigates the spatial retargeting problem, introducing the basic operators as well as the more advanced techniques that attempt to preserve image content and structure based on aspects of the human visual system. Chapter 6 discusses quality metrics for the evaluation of retargeting algorithms described in the previous chapters. In general, methodologies to compare output images produced by the retargeting techniques need to be developed. As the reader shall see, subjective quality assessment is a possibility; however, it is tedious and can only be applied on a limited set of images. Objective quality assessment is able to overcome these drawbacks, but it is limited by the current state of knowledge on how the human visual system and perception works. Both types of approaches are presented in Chapter 6 to provide an up-to-date overview to the reader.

Acknowledgements

I would like to thank Sumanta N. Pattanaik, Mateu Sbert, Karol Myszkowski, Attila and Laszlo Neumann, Yiorgos Chrysanthou, Efstathios Stavrakis as well as Werner Purgathofer for support in my research activities. Great thanks go to my fiancé Despo Ktoridou, who has supported me in all these years. I would like to thank, with all my heart, my mother Franca and grand-mother Nella, who are always in my mind. Grateful thanks to my father, Sincero, and brothers, Marco and Giancarlo; they have always supported my work. Every line of this book, and every second I spent in writing it, is dedicated to all of them. This work was partially supported by Ministry of Science and Innovation Subprogramme Ramon y Cajal RYC-2011-09372, TIN2013-47276-C6-1-R from the Spanish government, and 2014 SGR 1232 from the Catalan government.

Alessandro Artusi

I am heavily in debt for the support I have received from my parents, my brothers, my wonderful nieces, and my sisters in law. I also greatly thank my friends and colleagues. Now, to rest in full joy, I want to listen to the beautiful Dawn Song. Finally, I am very grateful to the outstanding co-authors whose hard work has made this book a reality. Writing a book and seeing it published is challenging as much as greatly rewarding. Domo Arigato!

Francesco Banterle

I'd like to thank my coauthors Karol Myszkowski, Rafal Mantiuk, Martin Cadik, Dawid Pajak, and Hans-Peter Seidel.

Tunç Ozan Aydın

We thank all our coauthors on spatial retargeting works and courses throughout the years, in particular YuShuen Wang, Ariel Shamir, Ofir Weber, Alexander Sorkine-Hornung, and Daniel Graf. This work was supported in part by the ERC Starting Grant iModel (StG-2012-306877).

Daniele Panozzo and Olga Sorkine-Hornung

1

Introduction

The human visual system (HVS) is a complex system that is capable, through several mechanisms, to transform the light striking the human eyes into meaningful information. Thanks to these mechanisms, we are able to adapt our visual system to different lighting conditions from moonlight to sunlight and can perceive a complex information such as colors. The main problem that we usually face, is the incapability to precisely reproduce this data in different media that may have different characteristics such as dynamic range, color spaces, color reproduction, etc. In this introductory chapter, the basic concepts of how the HVS works will be introduced; followed by an overview of color and the extended luminance dynamic range. These concepts are essential for understanding much of this book.

1.1 Human Visual System

The term *dynamic range* is used to describe the ratio between the largest and smallest possible values of a measurable quantity. When we are considering light energy, its dynamic range is the ratio between the highest and the lowest light energy values. In the real world, this range can be vast i.e, from starlight to sunlight, and the HVS is capable of coping with these lighting conditions through a complex process called *adaptation*. This process involves several mechanisms and some of them have been largely studied and used in many areas. However, many other mechanisms are still not fully understood today.

If we want to simulate and reproduce what a human being sees and perceives, we have to take into account *visual adaptation*. Here, we will introduce the basic components of the HVS and its adaptation process.

1.1.1 The Eye

The eye has approximately a spherical shape [295] and is the fundamental part of the HVS. Its structure is shown in Figure 1.1. The radius of the eye is of about 12 mm; its movements in its bony orbit are due to the use of three types of muscles. The eye consists of several components such as:

cornea, pupil, lens, aqueous humor, vitreous body, and retina. We will introduce the fundamental ones.

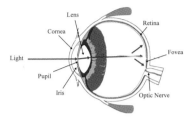

Figure 1.1. Human eye structure.

The external stimulus light, when striking the eye, encounters the first eye layer, the *cornea*. This has a complex lamellar structure and it is transparent without blood vessels. The hole in the iris diaphragm is where the *pupil* is located, and where light passes through. The *lens* is a biconvex multi-layered structure. The shape of the lens changes through accommodation, and these changes occur mostly at its anterior surface, which touches the iris. The *aqueous humor* is a liquid located between the cornea and lens, and it maintains the structural integrity of the eye as its main function [295].

The *retina* is a complex and multi-layered structure [295]. There are two elements of particular importance in the retina: *fovea* and *photoreceptors*. The fovea is a particular area of the retina where vision is more acute. There are two types of photoreceptors in the retina: *cones* and *rods*. The cone photoreceptors are also present in the fovea, but they have a different structure than the cones in the retinal region. In the fovea, there is the highest density of cones, which gives to the fovea its exceptional capacity to resolve the fine details in an optical image focused there. The most central cones and each group of rods have a direct connection to the brain through the inner surface of the retina and the optic nerve. In addition to these direct connections, there are myriad local cross connections in the retina [133]. The role of the cones and rods in the adaptation process (used by the HVS) to cope with the vast luminance dynamic range available in the real world, will be extensively described in Section 1.1.2.

An interesting fact to note is that the focusing process of the rays of light striking the cornea happens in the retina and lens. However, the iris diaphragm regulates the pupil size and consequently the amount of light entering in the eye. All these processes regulate the sharpness of the image perceived by the HVS [133].

The radiant energy associated with the light is converted to nerve activity by the pigments inside the rods and the cones [133]. These patterns of nerve activity are propagated from the receptors to the brain, and here this information is processed giving human beings the capacity to finally see the image.

1.1.2 Visual Adaptation

Changes in the environmental illuminant conditions, where the eyes are exposed, activate a process called *visual adaptation*. In this process, the HVS gradually adjusts itself to these changes [133]. A typical example of this process, experienced every day, is when we move from an environment with a low luminance level to an environment with a much higher luminance condition; e.g. coming out of a tunnel in a bright day. In this case, we all have experienced an initial level of visual discomfort, where we have reduced vision capabilities; i.e, the glaring effect. This first step is then followed by a process that gradually allows our vision system to restore normal vision.

The visual adaptation is a complex process that is achieved through the combination of mechanical, photochemical, and neural processes which are happening inside in the visual system [133]. They generate physiological phenomena that play a role in visual adaptation [88].

These mechanisms allow extending the limited response range of the neural *units*, that is, of only about *1.5 log units*, allowing the HVS to work over a luminance range of nearly *14 log units* [254]. In other words, through these mechanisms, the HVS moderates the effects of changing levels of illumination on visual response and provides sensitivity over a wide range of ambient light levels [88].

- **Saturation Process**: As any physical system, the HVS is also affected by a saturation process, which is due to the limited photoreceptor cell response to light at a maximum rate. This cell's response is generated by the chemical reactions produced by the light striking the cell's photo pigments [88]. The saturation process happens when two conditions are reached. First, these chemical reactions are near their maximum level. Second, there is an increase in the amount of light striking the photo pigments. As a result, cells are not able to signal the increase of the amount of light. In other words, during the saturation process, we experience small changes in the cell response to the light increment, when compared to the cell's response in normal conditions.

- **Visual Threshold vs. Intensity**: A visual threshold is defined by the probability p of seeing a visual difference in any visual attributes exhibited by a given set of stimuli [295]. In case of luminance intensity, Figure 1.2 shows the results of a threshold experiment that measured the changes in visibility that occur with changes in the level of luminance. These curves are known as *Threshold vs. Intensity*

(*TVI*) functions, and they show the relationship between the threshold and the background luminance's L. This relationship is described by *Weber's law*. This defines a linear behavior on a log-log scale. Weber's law is expressed as $\Delta L = kL$, and it indicates that the visual system has constant contrast sensitivity since the Weber contrast, $\Delta L/L = k$, is constant over this range [212].

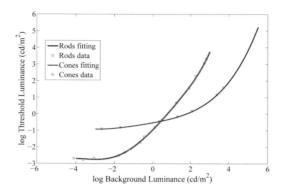

Figure 1.2. A psychophysical model of detection thresholds over the full range of vision (*TVI*).

- **Contrast Sensitivity**: Figure 1.3 shows the spectral sensitivities of the photoreceptor rod and cone systems at different luminance levels. They are described by the *scotopic, mesopic* and *photopic* luminous efficiency functions. At the scotopic level, Figure 1.3 (a), the sensitivity of the rod system is much higher than the sensitivity of the cone system. This suggests that the rod system is active. Since it is achromatic, at this level of illumination, color vision is not activated yet. At the mesopic level, the rods and cones systems are nearly equal in absolute sensitivity; see Figure 1.3 (b). Therefore, both systems are active and colors start to be perceived. At the photopic level, Figure 1.3 (c), the cone system is dominant and the absolute sensitivity of the rod system has dropped. In this case, the color vision is fully active [88].

1.1.3   Visual Adaptation Types

The most important types of adaptation, which we experience every day, are *dark, light*, and *chromatic* visual adaptation. Dark and light adaptation refer to the adjustment of the visual mechanism to changes in the rate at which radiant energy enters the eye. In contrast, the chromatic adaptation

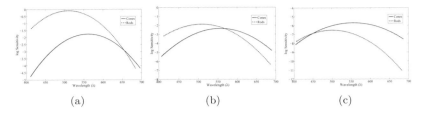

(a)   (b)   (c)

Figure 1.3. Changes in the spectral sensitivity of the HVS at (a) scotopic, (b) mesopic, and (c) photopic illumination levels.

refers primarily to the adjustment of the visual mechanism to change in its spectral distribution [133].

More importantly, the adaptation process does not happen instantaneously. As described in Section 1.1.2, this process gradually happens, so time plays an important role in it. Another aspect to notice is that the adaptation time is not the same for all types of visual adaptation. The HVS requires more time to adapt itself to low illumination (dark) than to adapt itself to high illumination (light) levels.

- **Dark Visual Adaptation**: Dark visual adaptation happens when luminance levels decrease. This is typical when we rapidly move from a photopic to a scotopic level of illumination, and we experience temporary blindness followed by an increase of the visual sensitivity that helps the HVS to gradually adapt its capability to visually perceive objects. During this adaptation process, the cones are gradually deactivated and colors are not perceived. The time required to realize the dark visual adaptation was measured by Hecht [112], and these results are shown in Figure 1.4. The cone and rod systems behave differently. Less adaptation time is required for cone than for rod, and the visual system is completely adapted after 35 minutes.

- **Light Visual Adaptation**: Light adaptation is experienced when we pass from scotopic to photopic, with a consequent increase in the overall level of illumination. This results in a decrease in visual sensitivity of the HVS and gradual activation of the cone system and deactivation of the rod system [78]. The behavior of the rod and cone systems, for light visual adaptation, is shown in Figure 1.5. The graphs show that light adaptation of the rod system is extremely rapid in the scotopic range; it needs only 2 seconds. In contrast, for the cone system, the light visual adaptation is much slower and it is on the order of minutes.

- **Chromatic Visual Adaptation**: Beside the visual adaptation for

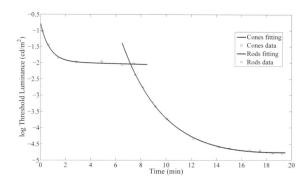

Figure 1.4. The time course of dark visual adaptation in the cone and rod systems.

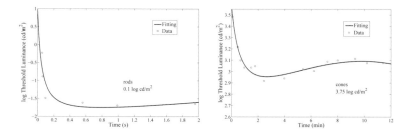

Figure 1.5. The time course of light visual adaptation in the cone and rod systems.

dark and light environmental lighting conditions, we have a very important adaptation mechanism called *chromatic visual adaptation*. This visual adaptation is characterized by the capability of the HVS to adjust itself to widely varying colors of illumination. The goal of this type of visual adaptation is to preserve the appearance of object colors [78].

1.1.4 Retinal Process

The mechanisms that control the time-dependent adaptation to varying luminance conditions occur inside the retina. The majority of the retinal cells can perceive only a small range of luminance values, compared to the entire luminance interval present in a scene. This range is continuously adjusted to adapt to light. Equation 1.1 below describes this process [213]:

$$R(I) = R_{\max} = \frac{I^n}{I^n + \sigma^n}, \tag{1.1}$$

where I is the light intensity, R is the neural response $R \in (0, R_{\max})$, the constant σ is the value I that causes the half-maximum neural response, and n is a sensitivity control.

1.2 Color

The color is the physiological sensation that the HVS produces when the eye is stroked by a light reflected, emitted, or transmitted by an object. In this section, we will analyze the nature of color and its attributes. Furthermore, we will describe how color can be represented in a three-dimensional space, or color space, and how color can be managed to deliver the desired characteristics.

1.2.1 What Is Color?

Color is often identified as a characteristic of a given object. This is, however, not correct. Color can be defined as a result of our physiological perception. This depends on several factors such as material properties, conditions of the observer, characteristics of the visual system, and neural processes. Given the followings definitions:

- The *stimulus* is the visible radiation that strikes the eye;

- The *stimulus object* is the source that generates the stimulus;

- The *color answer* is the response of the HVS to the stimulus;

we can say that in colorimetry the color is associated with the stimulus. On the other hand, the perception of color is defined as the color sensation, and the term *color* is used to identify a characteristic of the stimulus.

The attributes that are perceived when a color is observed are *hue*, *saturation*, and *lightness*. The hue is defined as the attribute of color by means of which a color is perceived to be red, yellow, green, blue, purple, etc. [78]. For example, pure white, black, and grays do not have hue, and they are called *achromatic* colors. On the contrary, colors that have hue are called *chromatic* colors. To define what saturation and lightness are, we first need to introduce the concept of *brightness*. Brightness [78] is defined as the perceived *luminance*, where luminance is the intensity of light emitted from a surface per unit area in a given direction. Saturation is defined as the colorfulness of an area judged in proportion to its brightness [78]. Lightness, as defined by Fairchild [78], is the brightness of an area judged relative to the brightness of a similarly illuminated area that appears to be white or highly transmitting.
How is a color reproduced?

The color can be produced combining or mixing three basic colors called *primaries*. Two approaches are typically used to reproduce colors: the *additive* one and the *subtractive* approach. The former is based on the concept that colors can be added to obtain a new one, see Figure 1.6 (a). This is also how the visual system mixes colors and CRT/LCD monitors work. The latter is based on the selective removal of wavelengths from light to produce a different color; see Figure 1.6 (b). Printers typically work with the subtractive approach. The primary colors are respectively referred to as additive primaries and subtractive primaries for the additive mixture and subtractive mixture techniques.

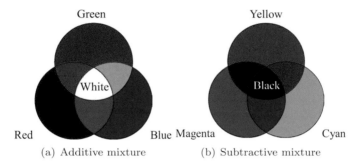

(a) Additive mixture   (b) Subtractive mixture

Figure 1.6. Typical approaches of color mixing.

1.2.2 Color Space

In real applications, colors need to be represented in numerical terms. This is the main goal of a science called *colorimetry*. In colorimetry, the authority is the Commission International de l'Éclairage (*CIE*), which is an international institute that is in charge of defining the standards and the procedures for using colors in practical applications.

In colorimetry, the perception of color from a stimulus of arbitrary spectral composition is described by three values called *tristimulus values* [89]. These values describe the position of a color in a three-dimensional space called the *color space*. In 1931, the CIE introduced a set of standard stimuli and the quantities of these stimuli to reproduce all colors of the visible spectrum by additive mixture. This data defines the Standard Colorimetric Observer 1931. Two standard observers have been introduced by the CIE, called 2- and 10-degree observers. These numbers identify the view angle of the field of view for an average observer. In other words, they describe the characteristics of an average observer with a specific visual field. The primaries that are used are three chromatic lights: red (700 nm), green (546.1 nm), and blue (435.8 nm). The color matching functions $\bar{r}(\lambda)$, $\bar{g}(\lambda)$

and $\bar{b}(\lambda)$ are the numerical descriptions of the chromatic responses of the observer; see Figure 1.7. They are used to obtain the tristimulus values of any color stimulus starting from its power spectral distribution $I(\lambda)$ integrated over the visible range (380 nm, 830 nm):

$$R = \int_{380}^{830} I(\lambda)\bar{r}(\lambda)d\lambda \quad G = \int_{380}^{830} I(\lambda)\bar{g}(\lambda)d\lambda \quad B = \int_{380}^{830} I(\lambda)\bar{b}(\lambda)d\lambda.$$

$$(1.2)$$

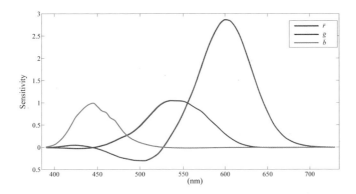

Figure 1.7. Color matching functions, $\bar{r}$, $\bar{g}$, and $\bar{b}$, of the CIE tristimulus values R, G, and B.

1.2.3 Color Space Categories

The color spaces can be divided into two main categories: *device dependent* and *device independent*. In device-dependent color spaces, the description of color information is related to the characteristics of a particular device (input or output). For example, in a monitor, it depends on the set of primary phosphors, while in an ink-jet printer, it depends on the set of primary inks. This means that a color (e.g. *R=250, G=20, B=150*) will appear differently when displayed on different monitors. On the contrary, a device-independent color space is not dependent on the characteristics of a particular device. This means that a color represented in this color space always corresponds to the same color information. A typical device-dependent color space is the *RGB* color space.

1.2.4 XYZ and xyY

For better manipulation of color information, the CIE defined a model where the colors are described by positive values, introducing the so-called imaginary primaries (tristimulus values): X, Y, and Z [89].

$$X = \int_{380}^{830} I(\lambda)\overline{x}(\lambda)d\lambda \quad Y = \int_{380}^{830} I(\lambda)\overline{y}(\lambda)d\lambda \quad Z = \int_{380}^{830} I(\lambda)\overline{z}(\lambda)d\lambda.$$
(1.3)

In the same way, color matching functions have been introduced for the new primaries $\overline{x}(\lambda)$, $\overline{y}(\lambda)$, and $\overline{z}(\lambda)$; see Figure 1.8. The chromaticity

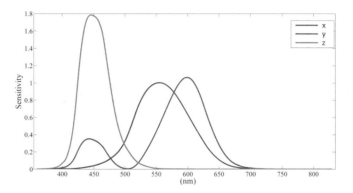

Figure 1.8. Color matching functions, $\overline{x}$, $\overline{y}$, and $\overline{z}$, of the new primaries X, Y, and Z.

coordinates are derived from the tristimulus values. They are the quantities related to the three primary stimuli needed to reproduce any color [1]. An important diagram can be derived from the chromaticity coordinates (see Figure 1.9) and it is called chromaticity diagram xy CIE 1931. A remarkable limitation of this diagram is its perceptual nonuniformity, which means that for color pairs represented by the same Euclidean distance, the perceived differences can be rather different. The CIE XYZ color space introduced by the CIE in 1931 is an example of a device-independent color space. All visible colors can be defined using only positive values. The Y value represents the luminance. A color that is defined in this system is refereed as Yxy. A third coordinate, z, can also be defined, but is redundant since $x + y + z = 1$ for all colors:

$$x = \frac{X}{X + Y + Z},$$
(1.4)

$$y = \frac{Y}{X + Y + Z}.$$
(1.5)

1.2.5   CIE $L^*a^*b^*$ and CIE $L^*u^*v^*$

Other device-independent color spaces are the CIE $L^*a^*b^*$ and CIE $L^*u^*v^*$. They have been introduced by the CIE to address the request to have color

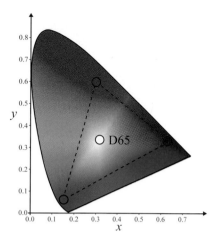

Figure 1.9. Chromaticity diagram xy CIE 1931.

spaces that are perceptually uniform. They are device-independent color spaces but suffer from being quite unintuitive despite the L^* parameter having a good correlation with lightness. Both color spaces are based on the CIE XYZ and are another attempt to linearize the perceptibility of unit vector color differences.

CIE L*a*b*

$$L^* = 116 f\left(\frac{Y}{Y_n}\right) - 16, \tag{1.6}$$

$$a^* = 500\left(f\left(\frac{X}{X_n}\right) - f\left(\frac{Y}{Y_n}\right)\right), \tag{1.7}$$

$$b^* = 200\left(f\left(\frac{Y}{Y_n}\right) - f\left(\frac{Z}{Z_n}\right)\right), \tag{1.8}$$

$$f(t) = \begin{cases} t^{\frac{1}{3}} & \text{if } t > 0.008856, \\ 7.787t + \frac{16}{116}, & \text{if } t \leq 0.008856. \end{cases} \tag{1.9}$$

The lightness value L^* is in the range $[0, 100]$ and (X_n, Y_n, Z_n) are the tristimulus values of the white point.

CIE L*u*v*

$$L^* = \begin{cases} 116(\frac{Y}{Y_n})^{\frac{1}{3}} - 16 & \text{if } \frac{Y}{Y_n} > 0.008856, \\ 903.3\frac{Y}{Y_n} & \text{if } \frac{Y}{Y_n} \leq 0.008856. \end{cases} \tag{1.10}$$

$$u^* = 13L(u' - u'_n), \tag{1.11}$$
$$v^* = 13L(v' - v'_n), \tag{1.12}$$

$$u' = \frac{4X}{X + 15Y + 3Z}, \tag{1.13}$$
$$v' = \frac{9Y}{X + 15Y + 3Z}. \tag{1.14}$$

Also in this case, L^* is in the range $[0, 100]$, (u'_n, v'_n) are the (u', v') chromaticity coordinates for the white point, and Y_n is the luminance of the white point.

1.2.6   RGB

The RGB color space is a Cartesian cube represented by the three additive primaries red, green, and blue. The gray-scale is the diagonal from black $(0, 0, 0)$ to white $(1, 1, 1)$, while the three color channels, RGB, are located on the three Cartesian axes. The RGB color space is used for describing color in several devices such as a display monitor, scanner, and digital camera.

The RGB color space is not a perceptually uniform color space, so it is not correlated with human visual perception, i.e., it is not linear with the visual perception response. Indeed, a variation of the same degree in the range $[0, 1]$ does not always produce the same variation of the perceived color. A linear relationship between the RGB and the XYZ color spaces exists. This allows us to define a simple linear transformation to convert a generic RGB color space to a device-independent XYZ color space such as

$$\begin{bmatrix} X \\ Y \\ Z \end{bmatrix} = \mathbf{M} \begin{bmatrix} R \\ G \\ B \end{bmatrix} \qquad \mathbf{M} = \begin{bmatrix} X_r & X_g & X_b \\ Y_r & Y_g & Y_b \\ Z_r & Z_g & Z_b \end{bmatrix}, \tag{1.15}$$

where the coefficients of the matrix $\mathbf{M}$ are the XYZ tristimulus values of the three RGB primaries. Typically, a gamma correction is required to simulate the nonlinear behavior of the display device. This step will be described in full detail in Chapter 2.

1.2.7   Color Management

A multimedia system usually includes several devices with different characteristics such as primaries, color space, technology used to reproduce colors, gamut, dynamic range, etc. This makes it difficult to convert a color reproduced by a specific device to a color that needs to be reproduced or visualized on another device. Color management can be seen as the methodology to overcome the faithful reproduction of colors between different devices. A typical color management system should take into account several aspects that characterize the differences between color devices:

- **Color Space**: Due to the specific technology used to reproduce colors, a color device encodes colors in a device-dependent color space. The limit of this, as explained in Section 1.2.3, is that represented colors in RGB cannot be linearly mapped to $CMYK$. Moreover, a digital representation of the same RGB color is differently reproduced on two different display devices. A solution to this issue is to convert the device-dependent color space into a device-independent color space (e.g. XYZ) allowing a unique representation of the same color. Starting from this new color representation, we have now a unique mapping to any device-dependent color space.

- **Color Gamut**: Color primaries may vary among color devices. Since the color primaries are defining the set of all colors reproducible by the device (*gamut*), we may have discrepancies between gamuts reproduced by different color devices. As as consequence of this, colors, that are reproducible on one device, may not be reproducible on another device; i.e., out-of-gamut colors.

- **Environment**: The environment where the colors are visualized plays an important role in how colors are perceived by the HVS; i.e., lighting conditions. Several appearance phenomena, generated by the environment where colors are observed, exist and can influence how colors are perceived. Many applications need to be able to predict and simulate these phenomena.

1.3   Extended Luminance Dynamic Range

Figure 1.10 shows how real-world luminance covers the three different luminance levels scotopic, mesopic, and photopic as described in Section 1.1.2. The rods are extremely sensitive to light and provide achromatic vision at

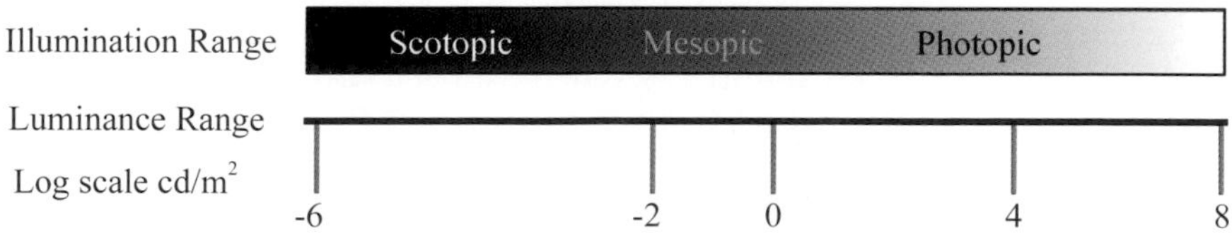

Figure 1.10. Real-world luminance.

scotopic levels of illumination ranging from 10^{-6} to 10^2 cd/m^2. The cones provide color vision at photopic levels of illumination in the range from 0.01 to 10^8 cd/m^2. At light levels from 0.01 to 10 cd/m^2, both the rod and cone systems are active [88].

1.3.1 High Dynamic Range Imaging

High dynamic range (*HDR*) imaging is a revolutionary imaging field; it provides the ability to capture, store, manipulate, and physically display real-world light values. These values can vary from starlight, 10^{-6} cd/m^2, to sunlight, 10^9 cd/m^2. The HDR imaging pipeline is presented in Figure 1.11 in all its stages.

In the first stage, *acquisition*, HDR content is captured from the real world. Different from classic imaging, there is the need to capture all light values; from starlight to sunlight. In order to achieve this, special imaging CCD/CMOS sensors are required, but these are extremely rare and expensive at the time of writing. Typically, HDR content is captured with traditional imaging equipment, where the scene to be captured is acquired multiple times by varying exposure time (ISO and iris diaphragm of the device can be varied as well) for capturing from very dark to very bright areas of the scene. Once different images of the same scene are acquired, they need to be post-processed in order to merge them into a radiance map; i.e., physical real-world light values. This step can be achieved with specialized software such as HDRShop [63], Adobe Photoshop [123], PictureNaut [184], etc. For a more in-depth review of these applications, we refer the readers to *The HDRI Handbook* [33]. Once HDR content is captured, we need to efficiently store it; the *storing* stage. This is because HDR content is stored using floating point values per color channel for coping with the great dynamic range that can be captured. Therefore, storing a raw picture requires four times the space of its counterpart at 8-bit per-color-channel; i.e., an HDR of a 2-megapixel picture requires 24 Mb without compression. To reduce the file size for HDR content, several efficient floating point representations have been proposed such as RGBE [286], LogLuv [153], half precision floating point [124], etc. However, these representations reduce the file size from 96 bits per pixel to 24 bits per pixel at best. In order to achieve greater compression rates, classic image and video compression algorithms have been extended. Regarding still images, JPEG 2000 and JPEG XR have been proposed to overcome the limited bit depth of the legacy coding system ISO/IEC 10918/T.81, widely known as the JPEG format. However, those standards have not been adopted by the digital photography market. As pointed out in [12], it is believed that an HDR image coding format should be backward compatible with the legacy JPEG format to facilitate its adoption and inclusion in

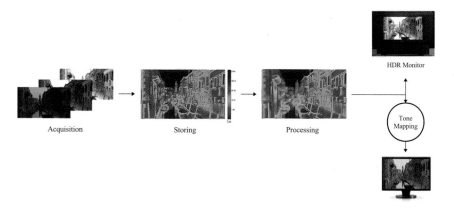

Figure 1.11. The HDR pipeline in all its stages. HDR content is captured, and is quantized, compressed, and stored. Further processing can be applied to it; e.g. filtering. Finally, the HDR content is tone mapped for SDR monitors or natively visualized using HDR monitors.

existing imaging ecosystems. Research works that have extended the JPEG format to handle HDR images exist: Ward and Simmons [285], Okuda and Adami [203] and Banterle et al. [26]. Recently, to address the lack of a coding standard for HDR images that is JPEG backward compatible, the JPEG Committee (ISO/IEC JTC1/SC29/WG1) released a new work, ISO/IEC 18477, also known as JPEG XT. This new standard is a two-layers codestreams design, where both layers are constrained to use legacy JPEG compression tools [12, 232]. Regarding videos, MPEG-4 and H.264 have also been extended in many works: Mantiuk et al. [177], Mantiuk et al. [176], Mai et al. [171]. In both cases, compression methods for still images and videos, the main idea is to reduce the HDR using sigmoids or logarithms (or custom functions), and to quantize content at 8 bits per-color-channel in order to fit the input stream of a traditional JPEG or MPEG encoder. Finally, the residuals between the original HDR and the 8-bit content are encoded in a separated layer.

At this stage, *processing*, we can manipulate the images and videos making use of operations from the classic imaging field. Once images or videos have been manipulated, we have two options to visualize their content. The first one is to use HDR monitors [244, 250], which natively display HDR content. The second option is to display HDR content on a standard dynamic range (*SDR*) monitor or TV. In this case, we need to retarget the information of the HDR image or video in order to fit traditional 8-bit per-color-channel devices; this process is called *tone mapping*. This is typically achieved by applying a function that reduces the luminance range. A typical tone mapping function is a sigmoid or a logarithm that can be applied per pixel, *global tone mapping*, or to a group of pixels, *local tone mapping*. The inverse problem exists when available 8-bit per- color-channel content needs to be retargeted to HDR monitors, i.e., *inverse tone mapping*. More details on tone mapping and its inverse will be presented in Chapter 2.

2

Tone and Color Retargeting

Faithful tone and color retargeting amongst different devices is of vital importance for many scientific and industrial applications. Today, a typical imaging system includes several devices that may differ in the color gamut and dynamic range that they can capture and either reproduce or visualize. An image can be captured with a digital camera and afterwards either visualized or printed with the typical result that its tone and colors are visibly different from the original captured data. Moreover, the technology limitations of the existing capturing and visualization devices may limit the quantity of digital information that we can capture and visualize. In the last years, a typical problem that has gained a lot of interest is the capture, manipulation, and visualization of the HDR luminance and the wide color gamut available in the real-world. Typically, HDR images are captured and then manipulated to preserve their tone and color information when visualized on traditional SDR displays. The advent of HDR images has introduced a series of complex issues that need to be addressed for using them in the traditional color retargeting pipeline for SDR images. This involves adapting the current technology that has been used for addressing the tone and color retargeting in the SDR imaging system. We will start by presenting the traditional color rendering pipeline for SDR images, and show how it can be adapted for HDR images. Existing techniques in the field will be presented and discussed covering all the steps of the extended color retargeting pipeline.

2.1 Color Retargeting Pipeline

Figure 2.1 shows a typical color retargeting pipeline adopted for a simple imaging system that uses SDR images. An image is first acquired with a digital camera, then it is visualized on a display. To faithfully reproduce the color of the digital camera image in the display, the following steps need to be considered:

- **Colorimetric Characterization**: The colorimetric characterization

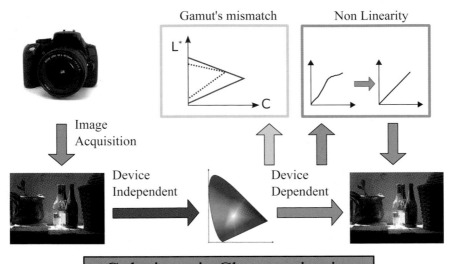

Figure 2.1. A typical color retargeting pipeline for SDR imaging system. An image is first acquired and represented in its device-dependent color space, i.e., RGB_c (digital camera). To have a unique representation of the same color, the RGB_c color space is converted to a device-independent color space, i.e., XYZ. To faithfully represent the digital camera color in the display system, the XYZ color space is converted to the device-dependent color space of the display RGB_d. Finally, the nonlinearity behavior of the display device and the gamut mismatches between the color gamuts of the two devices are taken into account.

step defines a mapping function between the device-dependent and device-independent color spaces and vice versa. Typically, three types of approaches are used to define this mapping function: physical models, exhaustive measurements, and empirical models. Physical approaches involve building mathematical models that find the relationship between the colorimetric coordinates and the digital signals of the device. The main advantage of these approaches is that they are robust, allowing for an easy recharacterization if some components of the imaging system are modified. Typically, few colorimetric measurements are required. The main disadvantage is that these models are often quite complex to derive and can be complicated to implement [15].

Exhaustive models, such as look-up tables (LUT), are used to define the colorimetric characterization mapping function via multidimensional interpolation. The main advantage of these approaches is that they do not require prior knowledge of how devices in the system work [14]. However, these approaches have several drawbacks. First, many measurements are needed. Second, the interpolation results may be not accurate, because data is highly nonlinear. Finally, the recharacterization of the color device maybe difficult when changes in the imaging system occur.

Empirical models collect a fairly large set of measured data and then statistically fit a relationship between device-dependent and device-independent colorimetric coordinates. In this case, the needed measurements are fewer than the ones needed for the LUT approach, but more than the ones needed for the physical models [15].

- **Gamut Mapping**: Mismatches between color gamuts reproduced by color devices may exist. In the example depicted in Figure 2.1, the source color gamut of the camera device may be larger than the destination color gamut of the display device. To be able to reproduce these colors on the display device, a naïve approach is to clip them to the gamut boundaries of the destination color gamut. However, this introduces strong desaturation, completely changing the appearance of these colors. Gamut mapping is adopted to help in reproducing these out-of-gamut colors while minimizing changes in their color appearance attributes.

- **Gamma Correction**: Color devices, such as a display, show a nonlinear behavior between the input voltage and its light intensity output. This nonlinearity is described through a function called the *electro-optical-transfer function* (*EOTF*). Several types of EOTFs have been proposed, and all of them define a common parameter γ for modeling this nonlinearity. To compensate for this nonlinearity, which allows you to predict the right light intensity value, a gamma correction step is required. This is performed through the application of an EOTF on the input digital data of the image to be visualized.

- **Color Appearance**: The viewing conditions, where a color image is observed, play an important role in how colors are perceived by the HVS. As partially described in Section 1.1.2, different viewing conditions are responsible for the activation of different types of adaptation mechanisms. As a result, the same color image under different viewing conditions may be perceived differently; i.e., have a different appearance. To take into account these phenomena in the color rendering pipeline, color appearance models have been introduced. They can predict changes of the so-called color appearance attributes under different viewing conditions to capture changes in how colors are perceived by the HVS.

2.2   Colorimetric Characterization

Color devices use a vendor-specific color space, where the physical representation of color is mapped to give a numerical representation (color space) that can then be used for any type of manipulation or visualization. This is mainly due to the following reasons:

- the type of technology used for reproducing colors in the device, i.e., additive or subtractive color mixing; type of color filter array in digital cameras and

- the type of primaries used in the color reproduction process, i.e., phosphors vs. LCD in displays, sensors and primaries of the color filter array in digital cameras.

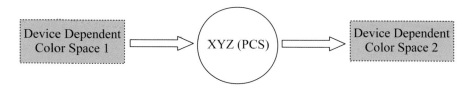

Figure 2.2. A typical colorimetric characterization process for an SDR imaging system. Starting from a device-dependent color space, a mapping to a device-independent color space and vice versa is defined.

Having different color spaces that depends from the technological characteristics of the device is a practical problem in an imaging system. Colors specified in a vendor-specific color space, also called a device-dependent color space, may be appear different when represented with the same color coordinates on a different device. To solve this problem, we need first to convert a device-dependent color space in a univocal representation where a color represented with three color coordinates is always identified as the same color. This means to construct a device-independent color space as defined in Chapter 1, which is called a *profile connection space* (*PCS*). This mapping process is called *colorimetric characterization* and a typical framework is shown in Figure 2.2. It defines a univocal mapping between a device-dependent color space and a device-independent color space and vice versa. The XYZ color space is typically used as a PCS. The colorimetric characterization process differs between different types of devices. In the following sections, we will introduce the colorimetric characterization of the digital camera (acquisition device) and of the display (visualization device), while the colorimetric characterization of the printer is out of the scope of this book.

2.2.1 Display

The colorimetric characterization of the display consists of two main steps:

- **First step**: Definition of the linear relationship between the RGB color space, characteristic of the display, and the PCS.

- **Second step**: This takes into account the display nonlinearity relationship between the input voltage and its light intensity.

In both cases, the inverse mapping is useful in any color management system. As shown in Chapter 1, the first step consists of the measurement of the primaries of the RGB color space expressed in their XYZ coordinates (XYZ color space is used as as PCS). To comply with technical and professional requirements, various RGB color spaces have been standardized. For example, ITU Rec. 709 [128] and ITU Rec. 2020 [130] are the two standard RGB color spaces defined for High-Definition TV (HDTV) and Ultra-High-Definition TV (UHDTV), respectively.

As an example here is the matrix conversion derived from their primaries, from RGB to XYZ, for ITU Rec. 709:

$$\mathbf{M}_{\text{Rec709}} = \begin{bmatrix} 0.4125 & 0.3576 & 0.1804 \\ 0.2127 & 0.7152 & 0.0722 \\ 0.0193 & 0.1192 & 0.9503 \end{bmatrix}, \tag{2.1}$$

where for ITU Rec. 2020 is:

$$\mathbf{M}_{\text{Rec2020}} = \begin{bmatrix} 0.636958 & 0.144617 & 0.168881 \\ 0.262700 & 0.677998 & 0.059302 \\ 0.000000 & 0.028073 & 1.060985 \end{bmatrix}. \tag{2.2}$$

The matrix conversion from XYZ to RGB, for ITU Rec. 709 is the following:

$$\mathbf{M}_{\text{Rec709}} = \begin{bmatrix} 3.2405 & -1.5371 & -0.4985 \\ -0.9693 & 1.8760 & 0.0416 \\ 0.0556 & -0.2040 & 1.0572, \end{bmatrix}, \tag{2.3}$$

where for ITU Rec. 2020 is:

$$\mathbf{M}_{\text{Rec2020}} = \begin{bmatrix} 1.716651 & -0.355671 & -0.253366 \\ -0.666684 & 1.616481 & 0.015768 \\ 0.017640 & -0.042771 & 0.942103. \end{bmatrix}. \tag{2.4}$$

Concerning the gamma correction step, several types of EOTF have been proposed and some of them standardized. They simulate different types of nonlinearity behavior of the display [29, 30].

- **Power Function**: This is the simplest EOTF type and it is the basis of the other more sophisticated EOTFs. The value d_i, $0.0 < d_i < 1.0$, is the digital value for the three $i = \{R, G, B\}$ color channels:

$$EOTF(d_i) = ad_i^{\gamma}, \qquad (2.5)$$

where a is called *gain* [29, 30], which controls the white level of the display, while γ is describing the nonlinear behavior of the system.

- **Gain-Offset-Gamma (GOG)**: In this model, an *offset* parameter $b = 1 - a$ is introduced to take into account the black level of the display:

$$EOTF(d_i) = (ad_i + b)^{\gamma}. \qquad (2.6)$$

- **Gain-Gamma-Offset (GGO)**: In this model, the γ value is applied only on the digital values d_i of the three RGB color channels:

$$EOTF(d_i) = ad_i^{\gamma} + b. \qquad (2.7)$$

- **Gain-Offset-Gamma-Offset (GOGO)**: In this model, the parameter c takes into account the internal *flare* of the display:

$$EOTF(d_i) = (ad_i + b)^{\gamma} + c. \qquad (2.8)$$

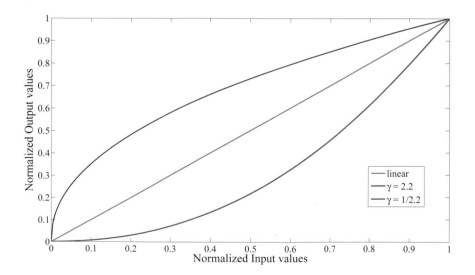

Figure 2.3. A typical power function EOTF and its inverse.

2.2.2 Digital Camera

The core parts of a digital single-lens reflex (DSLR) camera are the CCD sensors, were the main component is a photodiode that is responsible for converting the photon light into electrons (charge). This change in signal, due to the change in illumination, is called the camera response and, for a linear image sensor, is shown in Figure 2.4. The linearity of the response means that the charge is directly proportional to the amount of illumination striking the sensor. As shown in Figure 2.4, we observe two important phenomena. First, all DSLR cameras are affected by low light conditions, where information is lost. The camera sensor response to the illumination (charge) needs to rise above a certain level called the *noise floor*, before the sensor can distinguish between the noise and the illumination signal [264]. Second, at a higher level of illumination, the photodiode is unable to convert more photon light into electrons. This identifies the so-called saturation phase where all the information derived by the illumination above the saturation threshold is lost. Between these two levels of illumination, the noise floor and the saturation, we have the dynamic range that the camera is able to acquire. To have a large dynamic range, we need to quickly cross the noise floor and reach the point of saturation at the highest possible level of illumination [264]. To reproduce a color response that fulfills the

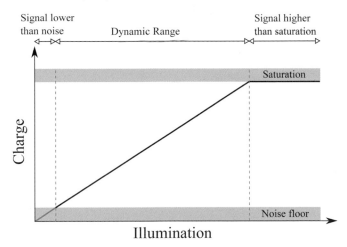

Figure 2.4. A typical CCD linear sensor response of a DSLR camera. The stored charge is directly proportional to the amount of illumination; after [264].

trichromatic model introduced in Chapter 1, RGB color filters are placed in front of the CCD sensors. Following the trichromatic model, we can define the RGB responses of the DSLR camera integrating, over the all visible spectrum, the product of the spectral power distribution of the light source

$I(\lambda)$, the reflectance/transmittance of the object $\rho(\lambda)$, and the responsivities of the color filters $D_r D_g D_b$:

$$R = \int_{380}^{830} I(\lambda) D_r(\lambda) \rho(\lambda) d\lambda, \qquad (2.9)$$

$$G = \int_{380}^{830} I(\lambda) D_g(\lambda) \rho(\lambda) d\lambda, \qquad (2.10)$$

$$B = \int_{380}^{830} I(\lambda) D_b(\lambda) \rho(\lambda) d\lambda. \qquad (2.11)$$

Recalling how the tristimulus values are calculated, we may notice that the difference between the RGB camera responses and the XYZ tristimulus values is in the different weighting functions used in their calculation. The tristimulus values XYZ, as defined in Chapter 1, use the CIE color matching functions $\bar{x}$, $\bar{y}$, and $\bar{z}$, which represent the responsivities of the HVS, and they have been measured through psychophysical experiments. Notice that in Equation 1.3 and Equation 1.2 the reflectance/transmittance of the object $\rho(\lambda)$ is included in the power spectral distribution $I(\lambda)$ term.

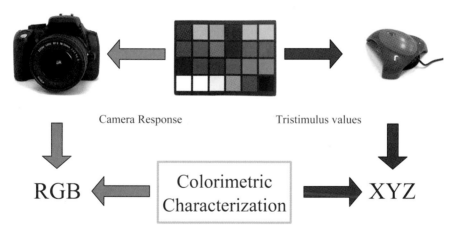

Camera Response   Tristimulus values

RGB ⟸ Colorimetric Characterization ⟹ XYZ

Figure 2.5. The workflow of the colorimetric characterization process for the DSLR camera. A set of colors (color checker) is used to define the mapping function between the RGB camera responses and the measured XYZ tristimulus values.

Colorimetric characterization of a DSLR camera follows the workflow shown in Figure 2.5. The idea is to have the RGB camera responses and the XYZ tristimulus values of the same number of color patches (color checker) used for the colorimetric characterization process. These are, respectively, the input and output of the mapping function that needs to be found, and this involves the solution of a classical linear system in the form of

$$C_{RGB} = \mathbf{A} \cdot C_{XYZ}, \qquad (2.12)$$

where C_{RGB} and C_{XYZ} are, respectively, the camera response and the measured tristimulus values. The matrix $\mathbf{A}$ represents the mapping function, and it can be found by solving the linear system.

2.3   Gamut Mapping

In an imaging system, the source and the destination gamuts are defined as the set of colors reproduced by the acquisition and the visualization devices, respectively, under specific viewing conditions. Note that a source gamut is the set of colors present in an image reproduced by the acquisition device. These two gamuts may be different due to differences in the color primaries used by the two devices, introducing the typical problem of gamut mismatch. Figure 2.6 shows two typical situations where either the mismatch is along the chroma component C (a), or both components (b) lightness L^* and chroma C. Gamut mapping solves these mismatches between color gamuts, mapping colors of the source gamut into the destination gamut to compensate for differences in the color gamut capability of the visualization device (destination gamut) [190].

2.3.1   Gamut Mapping Aims

The aim of a gamut mapping technique is to solve the possible mismatch between the source and destination gamuts, while maintaining, as much as possible, the matching of their color appearance attributes. However, it may happen that some colors are not physically reproducible by the visualization device. This makes it impractical to aim at matching the appearance attributes of individual colors in the image. Morovič has proposed that it is more advisable instead to aim for a match of the image's appearance attributes [191]. MacDonald [169] has identified a number of aims that are common to the majority of gamut mapping approaches that can be summarized as follows:

- **Preserving the gray axis of the image**: The black and white points of an input image (source gamut) are mapped to the black and white points of the destination gamut. This allows us to preserve the maximum luminance contrast of the input image.

- **Minimizing the hue shift**: Typically, to keep the hue of a reproduced color unmodified is a desirable feature.

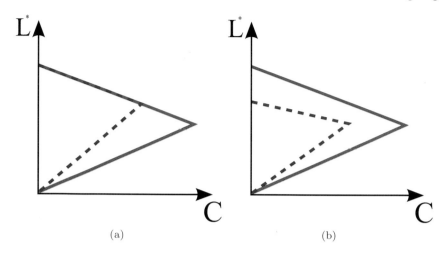

(a)   (b)

Figure 2.6. Gamut mapping problem. Two-dimensional representation of the gamuts of two devices: source (green) and destination (red). Mismatches between these two gamuts may appear: (a) the source gamut is outside of the destination gamut along the chroma component C; (b) the source gamut is outside of the destination gamut for both lightness L^* and chroma C.

- **Limiting out of gamut colors**: The aim of gamut mapping is to map the whole source gamut (image) into the destination color gamut (display). However, to eliminate extrema is desirable, because it allows a better overall appearance reproduction. This means adopting soft clipping to eliminate these extrema, and other approaches for the remaining colors.

- **Increase image saturation**: Since the destination gamut is limited in terms of saturation, it is desirable to use the available gamut to preserve chroma differences present in the source gamut (image).

- **Preserving color contrast**: The relative distance between chroma needs to be left unmodified, allowing the spatial relation between pixels to be maintained.

2.3.2   Rendering (Retargeting) Intent (RI)

The gamut mapping aim can be linked to the concept of rendering (retargeting) intent (RI) introduced by the International Color Consortium (ICC) [121]. The ICC has introduced four types of RI that define the intent (aim) of the color mapping between different gamuts (source and destination) in order to preserve the color appearance of the source gamut:

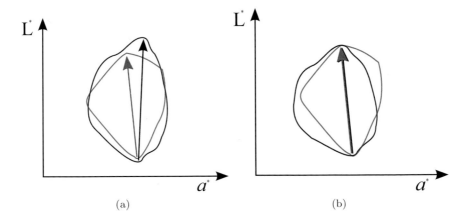

Figure 2.7. Gray axes alignment. The gray axis is the line that connects the black and the white points of the color gamut: (a) original source (red) and destination gamuts (blue) with the two gray axes not aligned; (b) after gray axes alignment.

- **Colorimetric**: The colorimetric intent defines two main intents: *relative* and *absolute*. The common scope of them is to preserve the relationship between in-gamut colors, while the mapping of out-of-gamut colors is not specified. However, the mapping of the out-of-gamut colors should be in-line with the intent and typically clipping is adopted.

  - **Relative**: The in-gamut colors are rescaled in order to have the white point of the source gamut mapped to the white point of the destination gamut. This intent is in-line with the aim of preserving the gray axis of the image in gamut mapping.

  - **Absolute**: The in-gamut colors are rescaled preserving the white point of the source gamut.

- **Saturation**: The scope of this intent is to preserve saturated colors.

- **Perceptual**: All colors of the source gamut are scaled into the destination gamut. Both in-gamut and out-of-gamut colors are affected. It differs from the colorimetric intent, because the out-of-gamut colors are not clipped but rescaled into the destination gamut with a more effective transformation. This intent is the closest to the aim of limiting the out-of-gamut colors.

2.3.3   Gamut Mapping Pipeline

A typical pipeline for a complete gamut mapping strategy is shown in Figure 2.8. The main goal of a gamut mapping technique is to minimize any distortion of color appearance attributes, such as hue and chroma, while minimizing mismatches between the source and destination gamuts. To accomplish this, a color space is required to be as close as possible to a perceptual uniform one. The perceptual uniformity of a color space is defined as the property of having a linear response between the mathematical measured distortion and how the distortion is perceived by the HVS in all areas of the color space. However, many color spaces are not perceptually uniform as shown by the MacAdams ellipses diagram (see Figure 2.8 (a)). The ellipses contain the colors that are not perceived as different, by an average observer, from the color located at the center of the ellipses. However, these ellipses should be circles, where all the contained color should have the same distance (measured distortion) from the center. This shows that color spaces such as the chromaticities diagram are not perceptually uniform so as they are not good candidates for computing distortion measures. The first step in a gamut mapping strategy is to select an appropriate color space with perceptual uniformity (see Section 2.3.4). Then, the gamut

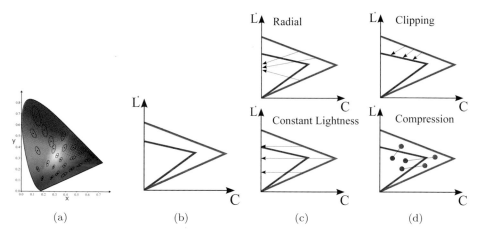

Figure 2.8. The four-step gamut mapping pipeline: (a) color space selection; (b) gamut boundaries computation for source (green) and destination (red) gamuts, e.g. slice of hue at 1 degree; (c) mapping directions, and (d) mapping approach selection. Examples of mapping directions are constant lightness lines and toward a gravity point (radial), e.g. average lightness, lightness corresponding to the maximum chroma of the source gamut (green). Examples of mapping approaches are clipping and compression.

boundaries descriptor (GBD) of the source and destination gamuts need to be computed. The output of this step is a 3D volume representation of the two gamuts. Figure 2.9 shows the gamut boundaries of two color gamuts, i.e., full sRGB vs. image gamut. The third and fourth steps consist of computing the intersection lines between the gamut boundaries and the mapping direction (gamut path), as well as choosing the mapping approach to be adopted.

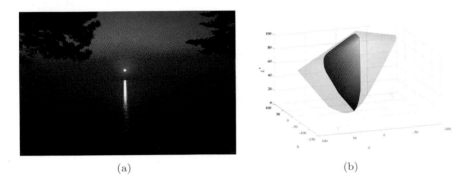

(a)   (b)

Figure 2.9. An example of visual representation of the color gamut: 3D volume gamut representation (b) of full sRGB color space (light colors) vs. the gamut of the image (a) (darker colors).

2.3.4 Color Space Issues

One of the major issues in gamut mapping, is the selection of an appropriate color space that is perceptually uniform as defined in Section 2.3.3. However, a perfect perceptually uniform color space has not been proposed yet. In a gamut mapping process, metrics calculating the distortions between color appearance attributes are used to achieve the mapping of colors from the source to the destination gamut. The use of a perceptually uniform color space can guarantee that if the mapping is done with minimal color appearance attributes distortion, they are perceived by the HVS as minimal.

One of the aims of gamut mapping is to minimize the hue shift of the source color gamut (see Section 2.3.1) while mapping them to the destination gamut. In practice, a gamut mapping technique is mapping the lightness and chroma of the source color gamut to the destination gamut at each fixed hue angle or following a line of constant hue metric [39]. If the color space is nonlinear, with respect to the hue, the perceived hue changes [39]. This is a typical problem found in the CIE $L^*a^*b^*$ color space, where blue and red regions are nonlinear (perceptually nonuniform) so in these regions, gamut mapping techniques may fail in minimizing hue distortion. A hue-linearized

CIE $L^*a^*b^*$color space exists [39], but it does not work well as the original CIE $L^*a^*b^*$color space outside of blue and red regions.

2.3.5 Point-wise Techniques

Point-wise techniques are per-pixel-based approaches, where the source gamut pixels are mapped to the destination gamut without taking into account any local spatial information about the pixels. As a result of this, a source gamut pixel is always mapped to the same destination gamut position, despite its spatial position in the input image. These techniques can be classified into two approaches: clipping and compression.

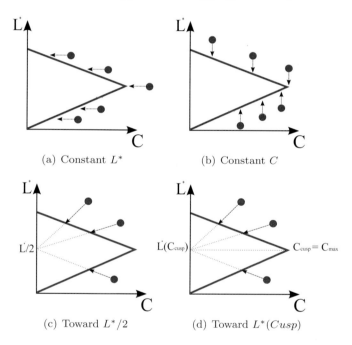

(a) Constant L^* (b) Constant C

(c) Toward $L^*/2$ (d) Toward $L^*(Cusp)$

Figure 2.10. Clipping approaches based on the mapping line directions and the color appearance attribute changes (lightness L^*, chroma C or both).

Clipping changes only the colors that are outside the destination gamut, mapping them on its gamut boundaries. The colors are untouched if they are located inside the destination gamut. The clipping approaches may differ from the gamut path used by the mapping step; see Figure 2.10:

- Horizontal clipping, see Figure 2.10 (a): Only chroma C is clipped toward horizontal lines with constant lightness L^*.

- Vertical clipping, see Figure 2.10 (b): Only lightness L^* is clipped toward vertical lines with constant chroma C.

- Radial clipping, see Figure 2.10 (c-d): Both chroma C and lightness L^* are clipped toward a center of gravity. The center of gravity lies on the lightness axis. This is defined either by a constant value such as its center $(L^*/2)$, or by a variable value such as the lightness corresponding to the maximum chroma value of the destination gamut (chroma cusp).

- Minimum color difference, ΔE, clipping: Each out-of-gamut color is mapped to the destination gamut surface location that minimizes the ΔE calculated in a specific color space. The ΔE defines the color difference between the color coordinates of the out-of-gamut pixel and its new location.

The major drawback of the clipping technique is shown in Figure 2.11. Out-of-gamut colors that belong to the same projection line are mapped to the same color of the gamut boundary of the destination gamut; see Figure 2.11 (a). This will clearly generate loss of local image details as shown in Figure 2.11 (c).

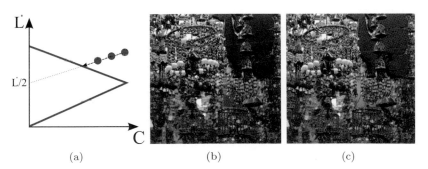

(a)   (b)   (c)

Figure 2.11. Clipping drawbacks: (a) Clipping may generate loss of local image details due to the fact that colors belonging to the same projection line (green) will be mapped to the same color within the boundaries of the destination gamut (red line). (b) An input image before clipping. (c) Image in (b) after clipping in the CIE $L^*a^*b^*$ color space; note that red colors have been clipped.

Compression changes all the colors of the source gamut to better fit into the destination gamut, while trying to preserve the matching between their color appearance attributes. Typical compression mapping functions are shown in Figure 2.12 and Figure 2.13, and they can be applied to both lightness, L^*, and chroma, C:

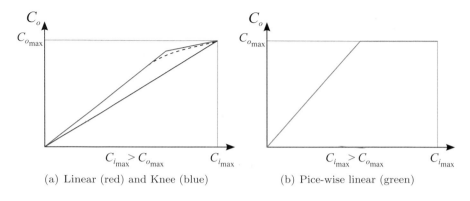

(a) Linear (red) and Knee (blue)   (b) Pice-wise linear (green)

Figure 2.12. Examples of compression mapping functions.

- Linear mapping linearly scales the source gamut range SG into the destination gamut range DG. Given the two extrema of the two ranges, $[S_{\min}, S_{\max}]$ and $[D_{\min}, D_{\max}]$, the mapping is performed as (Figure 2.12 (a) red line):

$$DG = D_{\min}\left(1 - \frac{SG - S_{\min}}{S_{\max} - S_{\min}}\right) + D_{\max}\left(\frac{SG - S_{\min}}{S_{\max} - S_{\min}}\right). \quad (2.13)$$

- Piece-wise linear offers a linear behavior for part of the source gamut, and for the remaining range clipping is applied ((Figure 2.12 (b) green line).

- Sigmoid function; a typical example is shown Figure 2.13, and it is defined as:

$$DG = \frac{1}{2\sigma}e^{-\frac{(SG-\mu)^2}{2\sigma^2}}, \quad (2.14)$$

where μ and σ control the shape of the sigmoid. μ moves the center of the sigmoid (Figure 2.13 (a)), while σ varies its slope (Figure 2.13 (b)).

- Knee-function is characterized as having a linear behavior for low-mid-range values, while for high-range values it offers a smoother solution than clipping (Figure 2.12 (a) (blue line)). As a result, the low-mid-range values of the source gamut are preserved, while the high-range values are softly compressed to the destination gamut.

Compression techniques are in general more capable of preserving details when compared with clipping techniques. However, due to the fact that spatial information is not used, local details may be lost. Another typical drawback of the compression techniques is to reduce saturation [36].

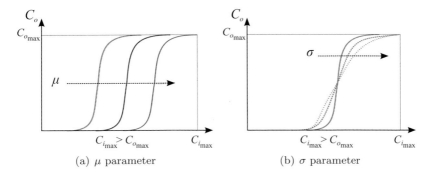

(a) μ parameter (b) σ parameter

Figure 2.13. An example of the sigmoid function on the chroma channel C^*. A sigmoid mapping function is defined with two parameters: μ, which moves the center of the sigmoid (a) and σ that modifies its slope (b).

2.3.6 Spatial Techniques

Spatial techniques aim to preserve the spatial relationships between neighborhood pixels. This allows us to better reproduce fine details while preserving edges, when compared to point-wise techniques. As result of this, two identical color pixels, belonging to a different spatial locations in the image, may be mapped to two different color pixels based on the information carried by their neighbor pixels. Bonnier et al. [36] have extended the classical gamut mapping aims, as specified in Section 2.3.1, to include two other important aims. The first is to preserve the spatial information, which is achieved by preserving the color relationship between neighbor pixels. The second aim is to avoid a preserving spatial relationship technique that introduces artifacts into the output image. These spatial techniques can either use models of the HVS to minimize the differences between the input and output images, or separate the high and low frequencies of the input image and then compress the low frequency and reinsert the details contained in the high-frequency component of the input image [36].

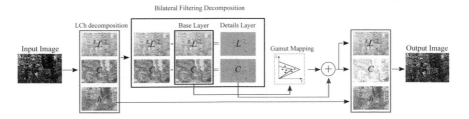

Figure 2.14. Spatial gamut mapping technique where the high and low frequencies information are separated through the use of the bilateral filter (see Appendix A).

Figure 2.14 shows a typical example belonging to the second category. A bilateral filter, see Appendix A, is used to decompose the input image in two scales: high- and low-frequency scales. The former is carrying the fine spatial details, while the latter contains large gradient information. Having the input image represented in a perceptually uniform color space CIE $L^*C^*h^*(I_{LCh})$, a bilateral filter can be separately applied to the lightness L^* and chroma C components, leaving the hue h untouched. The output of this operation is a low-frequency version of the lightness L_l^* and chroma C_l components, respectively. The high-frequency components (L_h^* and C_h) are obtained either through a difference or division between the input lightness L^* and chroma C and their low-frequency components L_l^* and C_l, respectively:

$$I(LCh)_h = \left(L^* - L_l^*, C - C_l, 0\right). \tag{2.15}$$

Once the two scale decomposition is performed for the lightness and chroma components, they can be separately manipulated for mapping the gamut of the input image into the gamut of a specific display. For achieving this mapping, any of the point-wise techniques and their combinations can be used. The final step is to merge the high- and low-frequency components to reconstruct the output image, leaving untouched the original hue h:

$$I(LCh) = \left(L_h^* + L_l^*, C_h + C_l, h\right). \tag{2.16}$$

2.4 Color Appearance Models

How the HVS perceives a color is influenced by the viewing conditions under which that color is observed. The viewing conditions are defined as the lighting under which a color is observed, the medium through which a color is observed, the background and the surround of the colors under observation, and the observer itself. Figure 2.15 shows the correlation between the color pixel location and its background and surround. All these factors are responsible for a series of phenomena called *color appearance phenomena*, which influence the color perception. In other words, if we have two color samples, expressed in tristimulus values as XYZ_1 and XYZ_2 with $XYZ_1 = XYZ_2$, and they look different, then a change in viewing conditions applies. In Section 1.2 we introduced a series of color appearance attributes such as brightness, lightness, colorfulness, saturation, chroma, and hue. The computation of these color appearance attributes is necessary for predicting the color appearance phenomena. Color appearance models have been developed to predict and simulate changes in the color appearance attributes, and consequently, how a color is perceived under different viewing conditions.

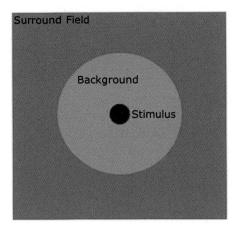

Figure 2.15. Color pixel position (stimulus - red) and its correlation with its background and surround fields. The stimulus is the area for which the color appearance is desired (taken at a 2-degree angle) (CIE 1931). The proximal field is the area around stimuli extending out another 2-degree, the background is the 10-degree field and the surround field is the area around the background.

2.4.1 Color Appearance Framework

A common framework can be derived for all color appearance models, as shown in Figure 2.16 [141]. Four common steps are implemented to predict the color appearance attributes. Starting from the tristimulus values XYZ, the first step is to account for the chromatic adaptation process. The second step is to convert the chromatic-adapted XYZ_c tristimulus values into the cone-adapted responses $L_aM_aS_a$. The third step is to separate the achromatic, A, and chromatic information or opponent responses, a and b, through a color decorrelation step. Finally, the color appearance attributes are predicted from the achromatic and chromatic information. The color spaces introduced in Chapter 1 are examples of simple non-

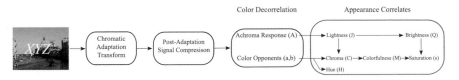

Figure 2.16. A general color appearance framework. All the color appearance models in literature includes these four steps.

spatial color appearance models introduced by the CIE. Based on this initial work, the CIE has started to extend the traditional colorimetry of these color spaces to predict the appearance of color observed under specific viewing conditions [54]. The result of this work is the standard color appearance model CIECAM97 and its extension CIECAM02. These color appearance models do not take into account the information of neighborhood pixels (spatial information). Both color appearance models require input parameters that specify the viewing conditions (surround) where the color sample is observed. These are the impact of surround c, the chromatic induction factor N_c, and the degree of adaptation F. The color appearance model CIECAM97 requires an extra input parameter, lightness contrast factor F_{LL}. These input parameters are defined in the standards [54, 189] and are used in the different steps.

2.4.2   Chromatic Adaptation Transform

Chromatic adaptation simulates the adaptation mechanism of the HVS to discount the color of the illumination. Image capturing devices are unable to account for this phenomenon, so they are incapable of adapting to an illumination source. Colors may appear different when the source illumination differs from the one used during the capturing process. Being able to simulate the discount of the color of the illumination source, allows us to overcome the above limitation of image capturing devices. The chromatic adaptation step consists of a chromatic adaptation transform (CAT) that predicts the new tristimulus values XYZ_c under a specific illumination source. As shown in Figure 2.17, the CAT follows a generic framework, where first the tristimulus values XYZ are converted to the cone responses LMS (matrix $\mathbf{M}$). Then the LMS cone responses are adapted to the new illumination source (white point) LMS_c (matrices $\mathbf{C}_1$ and $\mathbf{C}_2$). Finally, the adapted cone responses LMS_c are converted back to the new tristimulus values XYZ_c:

$$
\begin{bmatrix} L \\ M \\ S \end{bmatrix} = \mathbf{M} \begin{bmatrix} X \\ Y \\ Z \end{bmatrix}, \quad \begin{bmatrix} L_c \\ M_c \\ S_c \end{bmatrix} = (\mathbf{C}_2\mathbf{C}_1) \begin{bmatrix} L \\ M \\ S \end{bmatrix}, \quad \begin{bmatrix} X_c \\ Y_c \\ Z_c \end{bmatrix} = \mathbf{M}^{-1} \begin{bmatrix} L_c \\ M_c \\ S_c \end{bmatrix}.
$$
$$(2.17)$$

The differences between the various color appearance models lie in the coefficients of the matrix M, as well as in how the adaptation on the new illuminant source of the cone responses LMS is performed (matrices $\mathbf{C}_1$ and $\mathbf{C}_2$). The $\mathbf{M}$ matrices adopted by the CIECAM97 and CIECAM02 are given respectively by

$$
\mathbf{M}_{\mathrm{CAM97}} = \begin{bmatrix} 0.8951 & 0.2664 & -0.1614 \\ -0.7502 & 1.7135 & 0.0367 \\ 0.0389 & -0.0685 & 1.0296 \end{bmatrix}, \qquad (2.18)
$$

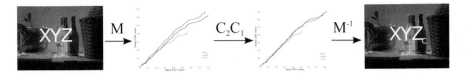

Figure 2.17. Chromatic adaptation transform: starting from the tristimulus values XYZ under the viewing conditions $\mathbf{C}_1$, the new tristimulus values XYZ_c, under the viewing conditions $\mathbf{C}_2$, are computed.

and

$$\mathbf{M}_{\text{CAT02}} = \begin{bmatrix} 0.7328 & 0.4296 & -0.1624 \\ -0.7036 & 1.6975 & 0.0061 \\ 0.0030 & 0.0136 & 0.9834 \end{bmatrix}. \tag{2.19}$$

Differences between these two color appearance models are also in the adaptation process of the LMS values to the new LMS_c values, under the new illuminant. The CIECAM97 adopts

$$\begin{bmatrix} L_c \\ M_c \\ S_c \end{bmatrix} = \begin{bmatrix} (D(\frac{1.0}{L_w}) + 1.0 - D)L \\ (D(\frac{1.0}{M_w}) + 1.0 - D)M \\ (D(\frac{1.0}{S_w^p}) + 1.0 - D)|S|^p \end{bmatrix}, \tag{2.20}$$

where L_w, M_w, and S_w are the LMS values for the reference white. Exponential nonlinearity is adopted for the short-wavelength S_c. The parameter p that models this nonlinearity behavior is given by

$$p = S_w^{0.0834}. \tag{2.21}$$

In Equation 2.20, D specifies the degree of adaptation (0.0 no adaptation, 1.0 fully adapted):

$$D = F - \frac{F}{[1.0 + 2.0(L_a^{\frac{1.0}{4.0}}) + \frac{L_a^{2.0}}{300}]}. \tag{2.22}$$

Similar equations are used for the CIECAM02 color appearance model:

$$\begin{bmatrix} L_c \\ M_c \\ S_c \end{bmatrix} = \begin{bmatrix} (D(\frac{Y_w}{L_w}) + 1.0 - D)L \\ (D(\frac{Y_w}{M_w}) + 1.0 - D)M \\ (D(\frac{Y_w}{S_w}) + 1.0 - D)S \end{bmatrix}, \tag{2.23}$$

where Y_w is the luminance of the reference white, and as above for Equation 2.20, L_w, M_w, and S_w are the LMS values for the reference white. Here, the exponential nonlinearity for the short-wavelength S_c is not used and D is computed as

$$D = F\left(1.0 - \frac{1.0}{3.6}e^{-(\frac{L_a+42}{92})}\right), \tag{2.24}$$

where the surround value F, for both color appearance models, is derived from a specific table that defines the type of surround of the environment where the color is observed such as average, dim, and dark. Nevertheless, different values for the two color appearance models are used for the same type of environment surround. Finally, the adapted tristimulus values XYZ_c are computed applying the inverse of the matrix $\mathbf{M}$ as shown in Equation 2.17.

2.4.3   Post-Adaptation Signal Compression

This step consists of converting the adapted tristimulus values XYZ_c to the cone fundamental responses (Hunt–Pointer–Esteves color space). On one hand, the complexity of the color appearance model is increasing. On the other hand, it has been shown that the use of a color space that is closer to cone fundamentals allows a better prediction of the color appearance correlates [189]:

$$\begin{bmatrix} L' \\ M' \\ S' \end{bmatrix} = \mathbf{M}_H \begin{bmatrix} X_c \\ Y_c \\ Z_c \end{bmatrix}, \tag{2.25}$$

where the matrix M_H is

$$\mathbf{M}_H = \begin{bmatrix} 0.38971 & 0.68898 & -0.07868 \\ -0.22981 & 1.18340 & 0.04641 \\ 0.00000 & 0.00000 & 1.00000 \end{bmatrix}. \tag{2.26}$$

Starting from the Hunt–Pointer–Esteves color space $L'M'S'$, to compute the adapted cone responses $L_a M_a S_a$, the CIECAM97 uses a hyperbolic function [54]. Here is the equation for the L' component of the cone responses:

$$L_a = \frac{40(\frac{F_L L'}{100})^{0.73}}{[(\frac{F_L L'}{100})^{0.73} + 2.0]} + 1.0, \tag{2.27}$$

while the CIECAM02 uses the Michaelis–Merten equation:

$$L_a = \frac{400(\frac{F_L L'}{100})^{0.42}}{[(\frac{F_L L'}{100})^{0.42} + 27.13]} + 0.1, \tag{2.28}$$

where F_L is computed as

$$F_L = 0.2k^4(5L_A) + 0.1(1 - k^4)^2(5L_A)^{\frac{1}{3}}, \tag{2.29}$$

$$k = \frac{1.0}{5L_A + 1.0}, \tag{2.30}$$

where L_A is the adapting luminance of the adapting field (see Figure 2.15) and it is computed as 20% of the white in the adapting field. The same equation is also applied to the M_a and S_a components for both color appearance models CIECAM97 and CIECAM02.

2.4.4   Color Decorrelation

Color decorrelation is used to separate the achromatic, A, from the color opponent responses, a and b. The color opponent responses correspond to the red-green and the yellow-blue opponent dimensions. The achromatic A response is computed, for the color appearance models CIECAM97 and CIECAM02 respectively, as:

$$A_{\text{CAM97}} = \left[2L_a + M_a + \left(\frac{1.0}{20} \right) S_a - 2.05 \right] N_{bb}, \qquad (2.31)$$

$$A_{\text{CAM02}} = \left[2L_a + M_a + \left(\frac{1.0}{20} \right) S_a - 0.305 \right] N_{bb}, \qquad (2.32)$$

where the noise term is a constant fixed to 2.05 for the CIECAM97 and to 0.305 for the CIECAM02 [189]. The N_{bb} parameter is the background brightness induction factor and it is computed as

$$N_{bb} = 0.725 \left(\frac{1.0}{n} \right)^{0.2}, \qquad (2.33)$$

where $n = Y_b/Y_w$ is the background induction factor, Y_b the luminance of the background, and Y_w the luminance of the reference white. The a and b color opponent values are computed, for both color appearance models, as

$$a = L_a - \frac{12.0 M_a}{11.0} + \frac{S_a}{11.0}, \qquad (2.34)$$

$$b = \frac{1}{9} [L_a + M_a - 2.0 S_a]. \qquad (2.35)$$

The hue angle h is also computed and used for color appearance correlates computation (see Section 2.4.5):

$$h = \tan^{-1} \left(\frac{b}{a} \right). \qquad (2.36)$$

2.4.5   Color Appearance Correlates

Once the achromatic and the color opponent responses are computed, the color appearance correlates can be derived. These are the lightness J, brightness Q, saturation s, chroma C, the colorfulness M, and the hue H. The lightness correlate J is computed from the achromatic responses as:

$$J = 100 (A/A_w)^{cz}, \qquad (2.37)$$

where A_w is the achromatic response for the reference white and c is an input constant to both color appearance models. It is specified in a given table as a function of the viewing conditions of the environment where the color patch is observed [189]. The z parameter is computed, for the color appearance models CIECAM97 and CIECAM02 respectively, as:

$$z_{\text{CAM97}} = 1 + F_{LL}\sqrt{n}, \tag{2.38}$$

$$z_{\text{CAM02}} = 1.48 + \sqrt{n}, \tag{2.39}$$

where n is defined as in Section 2.4.4. Since the brightness Q is related to lightness, as introduced in Chapter 1, it is computed, for the color appearance models CIECAM97 and CIECAM02 respectively, as:

$$Q_{\text{CAM97}} = (1.24/c)(J/100)^{0.67}(A_w + 3.0)^{0.9}, \tag{2.40}$$

$$Q_{\text{CAM02}} = (4.0/c)(\sqrt{J/100})(A_w + 4.0)F_L 0.25. \tag{2.41}$$

The colorfulness M, for both color appearance models, is computed as:

$$M = CF_L^{c_f}, \tag{2.42}$$

where c_f is equal to 0.25 for CIECAM02 and equal to 0.15 for CIECAM97, respectively. As introduced in Chapter 1, the saturation s is related to colorfulness M and brightness Q. The CIECAM97 computes the saturation, s, as a function of the color opponents and the cone responses as

$$s_{\text{CAM97}} = \frac{50\sqrt{(a^2 + b^2)}100e_{\text{CAM97}}(10/13)N_c N_{cb}}{L_a + M_a + (20/21)S_a}), \tag{2.43}$$

where N_{cb} is the chromatic brightness induction factor and it is computed like N_{bb}, and e_{97} is the eccentricity factor. Given an arbitrary hue angle h, it is computed via linear interpolation making use of the hue angle and eccentricity factor just below (h_i and e_i) and just above (h_{i+1} and e_{i+1}) the hue angle of interest:

$$e_{\text{CIECAM97}} = e_i + \frac{(e_{i+1} - e_i)(h - h_i)}{(h_{i+1} - h_i)}. \tag{2.44}$$

These values of hue and eccentricity refer to the primary colors as defined in the standards [54, 189]. CIECAM02 computes the saturation s as a function of the colorfulness M and the brightness Q as

$$s_{\text{CIECAM02}} = 100\sqrt{\frac{M}{Q}}, \tag{2.45}$$

where the chroma C is computed as

$$C_{\text{CIECAM97}} = 2.44 s_{\text{CIECAM97}}^{0.69} (J/100)^{0.67n}(1.64 - 0.29^n), \qquad (2.46)$$

$$C_{\text{CIECAM02}} = t^{0.9}\sqrt{J/100}(1.64 - 0.29^n)^{0.73}, \qquad (2.47)$$

where t for CIECAM02 is computed as:

$$t = \frac{e\sqrt{a^2 + b^2}}{L_a + M_a + (21/20)S_a}, \qquad (2.48)$$

$$e_{\text{CIECAM02}} = (12500/13)N_cN_{cb}[\cos(h\pi/180 + 2) + 3.8], \qquad (2.49)$$

where e_{CIECAM02} is the eccentricity factor for the CIECAM02. The hue H is computed via linear interpolation in am manner similar to Equation 2.44 for both color appearance models as

$$H = H_i \frac{100(h - h_i)/e_i}{(h - h_i)/e_i + (h_{i+1} - h)/e_{i+1}}. \qquad (2.50)$$

2.4.6 Spatial Models

The CIECAM97 and CIECAM02 color appearance models can predict several appearance phenomena, treating the color pixel of an image as completely independent stimuli. Spatial and temporal properties of the HVS are not included [76]. Models that include spatial properties are called *image color appearance models* or simply *spatial models*. *iCAM* is an example of a spatial model and its framework is depicted in Figure 2.18. As shown in Figure 2.18, iCAM takes as input the image and its low-pass

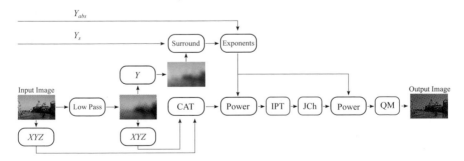

Figure 2.18. Spatial color appearance model: iCAM framework.

filter version (adapting stimulus) in relative XYZ tristimulus values. The absolute value of luminance Y_{abs} and of the surround Y_s are also required.

They are computed as a low-pass filtered version of the luminance stimulus of the input image, Y, at full resolution for Y_{abs}, and at lower resolution for Y_s. The first step, as in the non-spatial color appearance models, is to take into account the chromatic adaptation. This is achieved as in the CIECAM02 model by using the CAT02 model (see Section 2.4.2). This is given as input to Equation 2.23 to compute the chromatic adapted tristimulus values. However, $L_w M_w S_w$ values in Equation 2.23 are the LMS cone responses computed from the low-pass filtered input image. Once the chromatic adapted values $X_c Y_c Z_c$ are computed, we need to apply the postadaptation signal compression as in Equation 2.25 with $\mathbf{M}_H$:

$$\begin{bmatrix} L \\ M \\ S \end{bmatrix} = \begin{bmatrix} 0.4002 & 0.7075 & -0.0807 \\ -0.2280 & 1.1500 & 0.0612 \\ 0.0000 & 0.0000 & 0.9184 \end{bmatrix} \begin{bmatrix} X_c \\ Y_c \\ Z_c \end{bmatrix}. \qquad (2.51)$$

A power nonlinearity function is applied to the LMS cone responses:

$$L' = L^{0.43}, M' = M^{0.43}, S' = S^{0.43}. \qquad (2.52)$$

If the LMS values are negative, Equation 2.52 becomes

$$L' = -|L|^{0.43}, M' = -|M|^{0.43}, S' = -|S|^{0.43}. \qquad (2.53)$$

The final step is to separate the achromatic and chromatic components. This is done by introducing a new color space called IPT, where I refers to the achromatic response, while P and T are, respectively, the red-green and the yellow-blue color opponent responses:

$$\begin{bmatrix} I \\ P \\ T \end{bmatrix} = \begin{bmatrix} 0.4000 & 0.4000 & 0.2000 \\ 4.4550 & -4.8510 & 0.3960 \\ 0.8056 & 0.3572 & -1.1628 \end{bmatrix} \begin{bmatrix} L' \\ M' \\ S' \end{bmatrix}. \qquad (2.54)$$

The appearance correlates are simply derived from the IPT values. The lightness J is equal to the achromatic response I. The chroma, C, and the hue angle, h, are computed from the color opponent responses P and T as

$$C = \sqrt{P^2 + T^2}, \qquad (2.55)$$

$$h = \tan^{-1}\left(\frac{P}{T}\right). \qquad (2.56)$$

The brightness, Q, and colorfulness, M, are derived, respectively, from the lightness, J, and the chroma, C, by using the adaptation coefficient, F_L, which is computed as in the CIECAM02 model (see Equation 2.30):

$$Q = F_L^{\frac{1}{4}} J, \qquad (2.57)$$

$$M = F_L^{\frac{1}{4}} C. \qquad (2.58)$$

2.5 Color Retargeting Pipeline for HDR Imaging

HDR imaging aims to capture a wide dynamic range of luminance values and a wide color gamut to better convey the real-world user experience. Since this information cannot be stored in 8-bit per color channel, see Section 2.1; to fully represent this content a higher bit depth is required. Typically, 32-bit per color channel representation encoded in floating point is employed. Figure 2.19 shows an adapted color retargeting pipeline for SDR imaging (Section 2.1) to work with HDR values. In principle, many steps are in common with the color retargeting pipeline used for SDR imaging. However, these common steps require more complex processes due to the limitations of the available acquisition and visualization hardware.

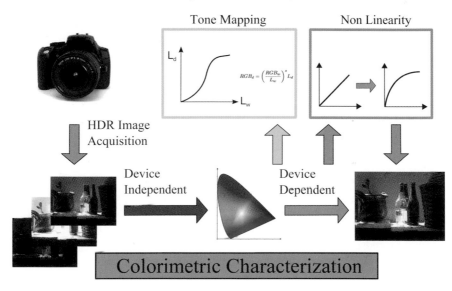

Figure 2.19. Color retargeting pipeline for HDR imaging. HDR capturing through the use of multi images taken at different exposure times, followed by colorimetric characterization and then the HDR content may be manipulated to be visualized on a display. The visualization process involves adapting the HDR content to the dynamic range available in the display, as well as the linearization process typical of the display device.

- **High Dynamic Range Acquisition**: HDR content can be captured in various ways; this depends on the available hardware technology. Typically, HDR images are acquired using SDR sensors (8/10-bit) by taking multiple images at different exposure times. Each SDR image in the sequence represents a slice of the full dynamic range available in the captured scene. The full dynamic range is recovered by merging the SDR images into a radiance map [64]. Existing colorimetric characterization techniques for SDR content (see Section 2.2) can be used for the colorimetric characterization of HDR content [140] [95]. This allows us to have the acquired data expressed in a device-independent color space.

- **Tone Mapping**: Mismatches between the acquired HDR content and the visualization devices are larger than the ones for SDR imaging. These can reach even 4 orders of magnitude for the luminance channel, and about of 2 orders of magnitude for the chroma channel. To visualize this information on SDR displays with a small color gamut, tone mapping operators (TMOs) are adopted. The process of compression is typically referred as *tone mapping*. TMOs mainly compress the luminance dynamic range maintaining as much as possible the contrast and details of the original input HDR image. Typically, they adopt heuristics to convey the original color information in the tone mapped image. Recently, works to retarget the original color information into the tone mapped image, have been proposed. They will be presented in Section 2.7.3 and Section 2.9. Tone mapping can sometimes be confused with gamut mapping. However, tone mapping substantially differs from gamut mapping, and the differences will be highlighted in Section 2.8.

- **Gamma Correction**: The current consumer display trends are providing toward features for HDR and a wider color gamut. A standard for a wider color gamut has been recently proposed (ITU-R Rec. BT.2020 [130]) and specifies a larger color gamut than the widely adopted color gamut standard ITU-R Rec. BT.709 [128]. On the other hand, we are in a transitory phase where displays that provide a different dynamic range and a wider color gamut are starting to appear on the market, e.g. 1,000 nits peak luminance. However, these displays still present the classical nonlinear behavior of the current SDR display. Therefore, the use of EOTFs is still required to simulate this nonlinearity.

- **Color Appearance**: As shown for the SDR color retargeting pipeline, colorimetric characterization (Section 2.2) and gamut mapping (Section 2.3) are not sufficient to convey the real color appearance of the visualized image. In the context of HDR imaging, tone mapping alone is not enough to faithfully reproduce the captured real-world appearance when visualizing it on a display. On the other hand, there is potentially a larger disconnection in terms of perception than what we have in the SDR color retargeting pipeline [7]. This makes the problem more difficult to solve.

2.6   High Dynamic Range Acquisition

Current consumer camera sensors (e.g. DSLR cameras, smartphones, tablets, etc.) are limited to capturing 8-bit per color channel images; if the device allows RAW writing, 12/14-bit per color channel images can be captured. This means that single shot HDR capturing is still not available. In fact, many scenes are challenging to acquire, and they require at least 16 bits per color channel; e.g. stained glass inside a dark church in the middle of a sunny day. In this section, we assume SDR images to be encoded using gamma encoding or a camera response function (CRF) and stored at 8 bits.

Typically, HDR images are acquired by taking photographs of the same

Figure 2.20. An example of capturing a scene at different exposure times; from left to the right: $\frac{1}{128}$ sec, $\frac{1}{32}$ sec, and $\frac{1}{8}$ sec.

scene at different exposure times, see Figure 2.20. There are essentially four main technologies for capturing photographs at different exposure times:

- **Single camera setup**

  - **Temporally varying global exposure**: This is the classic technology [64], where the global exposure time varies to capture the scene. This method works in optimal conditions when pictures are taken using a tripod and the scene is static. Otherwise, we can have misalignments if the camera moves, and/or *ghosts* if objects in the scene move. These issues can be partially solved using high-speed cameras with programmable shutter speeds [136], but their use is limited to bright scenes (more than 5,000 cd/m^2 on average), because images captured in very dark scenes (less than 10 cd/m^2 on average) can exhibit high noise levels (even more than 10 dB).

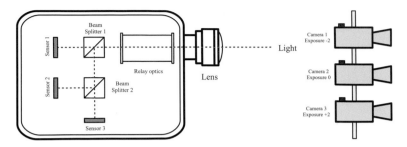

Figure 2.21. Different HDR acquisition techniques: On the left side is a diagram of a beam splitter setup; multiple SDR sensors in the same camera have different exposure times in order to capture most of the dynamic range of a scene. On the right side, the diagram of a multi-view setup; cameras can be arranged in a linear setup as in the figure or in a square setup; each camera has a different exposure time.

- **Spatially varying pixel exposure**: This technology [108, 195, 196] is essentially an extension of the widely used Bayer filters for color acquisition applied to exposure. The idea of these methods is to locally vary the exposure per pixel following a pattern. The exposure can be varied for each row or by groups of pixels in a squared tile. The main advantage of these techniques is that there is no need to apply image-alignment or deghosting techniques. However, they have disadvantages too. They inherit the same issues that Bayer filters have; i.e., they trade spatial resolution for dynamic range. Furthermore, this issue is exacerbated if a color Bayer filter is present in the camera setup as well. Therefore, high-quality camera reconstruction algorithms are the key to prevent deterioration of the final HDR image resolution.

- **Multi-cameras setups**

  - **Stereo or multi-view rigs**: This technology [34, 193, 273, 290] uses two or more cameras in a stereo or a multi-view setup [259], see Figure 2.21 (right). The advantage of multi-view approaches is the absence of ghosting artifacts. However, images, which are captured from different views, need to be aligned taking care of occlusions. Moreover, the issue becomes more exacerbated if the baseline is wide. This setup can be also expensive, because more cameras are required and a synchronization system has to be implemented to avoid temporal ghosts.

  - **Beam splitter setups**: This technology [151, 266, 279] is basically an extension of the 3 CCD techniques for high quality cameras and video-cameras (i.e., a CCD for each color channel to avoid Bayer filters and improve quality) applied to HDR; see Figure 2.21 (left). Basically, a camera is equipped with two or more sensors with different neutral density filters in front. They gather light arriving from a single lens through a beam splitter (prism), which splits light paths. This method has the advantage of creating a high-quality HDR image without alignment and ghosting issues. Nevertheless, high quality beam splitters are typically expensive, and the calibration process for correctly aligning light paths is time consuming.

Recently, Tursun et al. [271] presented an extensive and exhaustive review of the state-of-the-art techniques for aligning images and removing ghosts when capturing HDR images. These methods can be based on different image processing and computer vision algorithms [259] such as homography alignment through feature detections, optical flow methods, scene change methods, etc.

The ideal acquisition technology for HDR imaging could be a gradient camera [269] that captures gradients instead of intensities as in the common image sensors. This allows capturing of the most extreme HDR scene in a single shot without requiring exposure metering. Moreover, this technology can offer more advantages such as hiding quantization artifacts, hiding low-frequency noise, etc. However, it requires a Poisson solver in order to display an image, and a change in the manufacturing process of current image sensors. This latter issue is difficult to solve given the reluctance of image sensor and camera manufacturers in making extreme changes in their workflow.

After selecting an appropriate capturing technology, there is the need to capture the full dynamic range of a scene, from the darkest to the brightest regions of it. Fortunately, this process can be made automatic by employing exposure metering [91], which generates an HDR histogram out of SDR histograms captured from the viewfinder of a camera. From the HDR histogram, the exposure times for a specific scene are chosen to maximize the peak SNR, and modeled taking into account quantization noise, photon shot noise, as well as other signal-independent sources of noise.

Once all required SDR images are captured, they can be merged together into a radiance map [284]. This is achieved by summing all images scaled by the inverse values of their exposure times:

$$E(\mathbf{x}) = \frac{\sum_{i=1}^{n} \frac{1}{t_i} w(Z_i(\mathbf{x})) g(Z_i(\mathbf{x}))}{\sum_{i=1}^{n} w(Z_i(\mathbf{x}))}, \qquad (2.59)$$

where E is the recovered radiance at the $\mathbf{x}$ pixel location, g is the inverse camera response function, Z_i is the SDR of the image of a scene captured at the i-th exposure, t_i is the exposure time for Z_i, n is the number of images at different exposures, and w is a weighting function that removes outliers. w can have different shapes such as hat, tent, Gaussian, etc. w typically gives higher scores to well-exposed pixels, around 128, and lower scores to pixel at the end of the range, around 0 and 255. This is because outliers are usually overexposed (more than 250) or underexposed (less than 10) pixels. An example of a w function is the tent function, which is defined as

$$w(x) = \begin{cases} x - \tau_{\min} & \text{if } x \leq \frac{1}{2}(\tau_{\max} + \tau_{\min}), \\ \tau_{\max} - x & \text{if } x > \frac{1}{2}(\tau_{\max} + \tau_{\min}), \end{cases} \quad \text{with } x \in [0, 255]. \quad (2.60)$$

$\tau_{\min}$ and $\tau_{\max}$ are, respectively, parameters for defining the range of accepted values. They are typically set as $\tau_{\min} = 0$ and $\tau_{\max} = 255$.

Note that if g is unknown, it has to be estimated because not all cameras have the option to store values linearly (i.e., the RAW format). In this case, most of the methods assume g to be smooth and a monotonically increasing function. Mann and Picard [172] proposed a simple method for estimating g, which consists of fitting the pixel values at different exposures to a fixed CRF of the form $g(x) = ax^\gamma + b$. Although, this method is straightforward to implement and fast, it is not very accurate and does not support most real CRFs. Debevec and Malik [64] proposed a more general and robust method for recovering a CRF. Their method is based on properly rewriting the film reciprocity equation as

$$Z_i = g^{-1}(Et_i), \qquad (2.61)$$

$$g(Z_i) = (Et_i), \qquad (2.62)$$

$$\log g(Z_i) = \log E + \log t_i, \qquad (2.63)$$

therefore g can be recovered with an unknown E in a least-squared error sense. Given Equation 2.63, the following objective function needs to be minimized:

$$\mathcal{O} = \sum_{i=1}^{n} \sum_{j=1}^{M} \left(w(Z_i(\mathbf{x}_j)) \left[g(Z_i(\mathbf{x}_j)) - \log E(\mathbf{x}_j) - \log t_i \right] \right)^2$$

$$+ \lambda \sum_{x=T_{\min}+1}^{T_{\max}-1} (w(x)g''(x))^2, \qquad (2.64)$$

where M is the number of pixels used in the minimization (a subset of the whole image, e.g. 10% of pixels, must be used in order to quickly solve it and improve its stability), and $T_{\max}$ and $T_{\min}$ are, respectively, the maximum and minimum integer values in all images I_i. Note that the second term, starting from λ, is a smoothing term to ensure that g is smooth.

Debevec and Malik's seminal work has led to other different methods for recovering the CRF. For example, Mitsunaga and Nayar [187] proposed parameterizing g into a polynomial representation, and Robertson et al. [234] proposed a method that estimates the unknown response function as well as the irradiance, E, through the use of a maximum likelihood approach. Recently, these and other approaches have been investigated to understand their accuracy, robustness to noise, and consistency [207].

2.6.1 Colorimetric Characterization of HDR Content

As in the color rendering pipeline for SDR images, the idea is to have HDR content represented in a device-independent color space, i.e., XYZ, so it can be easily manipulated for visualization on any device. Since HDR content is acquired with DSLR cameras, the colorimetric characterization of HDR content consists of the conversion from the RGB color representation of the DSLR camera to the XYZ color space (PCS). Traditional colorimetric characterization models, as presented in Section 2.2, rely on defining a mapping function between these two color spaces where the RGB and the XYZ values are, respectively, the acquired scene and the measured tristimulus values of color patches of a color checker. This involves solving a linear system, as shown in Section 2.2.2, where the matrix $\mathbf{A}$, which defines this mapping function, is unknown.

A straightforward solution to the colorimetric characterization of HDR content is to directly employ the traditional colorimetric characterization models used for SDR images. This is typically employed when the HDR content is acquired with temporally varying global exposure techniques. In this case, each different exposed image, I_i, in the scene sequence is first converted into the linear $X_i Y_i Z_i$ color space (PCS), and then the reconstructed XYZ values of the HDR scene are computed as scaled and averaged values as [95]

$$X = t_n \frac{\sum_i X_i \frac{1}{t_i} w(X_i Y_i Z_i)}{\sum_i w(X_i Y_i Z_i)}, \qquad (2.65)$$

$$Y = t_n \frac{\sum_i Y_i \frac{1}{t_i} w(X_i Y_i Z_i)}{\sum_i w(X_i Y_i Z_i)}, \qquad (2.66)$$

$$Z = t_n \frac{\sum_i Z_i \frac{1}{t_i} w(X_i Y_i Z_i)}{\sum_i w(X_i Y_i Z_i)}, \qquad (2.67)$$

where t_i is the exposure value of the acquired image I_i of the scene sequence, t_n is a scale value to estimate the exposure value of the reconstructed HDR image, and w is a weight function. This emphasizes the contribution of pixels in the middle range while ignoring over/underexposed pixels, which may be noisy [95].

However, these traditional colorimetric characterization models rely on the following assumptions:

- $D50$ illuminant is the main illuminant used for the colorimetric characterization process when measuring the color checker patches with a spectrophotometer.

- The traditional color checker used in the colorimetric characterization process has limited dynamic range.

These assumptions can lead to the following drawbacks:

- The relative spectral power distribution, $I(\lambda)$, used in the calculation of the tristimulus values XYZ and the color values RGB, is normalized to 100 cd/m^2. This discharges the intensity scale of the illumination, making difficult to calibrate absolute values [140].

- The illuminant used to measure the reflectance of color patches with a spectrophotometer is different from the illuminant used to illuminate the acquired scene. This makes the traditional colorimetric characterization models for DSLR cameras ineffective when a different illuminant, from the one used in the scene acquisition is used. This requires repeating the colorimetric characterization of the DSLR cameras if the illuminant changes [140].

As shown by Kim et al. [140], the development of a transparent color checker backlit with a specific illuminant solves these drawbacks. First, the dynamic range of the color checker is not limited to the range $[0, 100]$ cd/m^2. Second, a spectroradiometer can be used for the measurements of the color checker patches, allowing the use of the same illuminant for the measurements and for the acquired scene. This makes the workflow for colorimetric characterization of HDR content identical to the workflow used for SDR images. Moreover, there is the advantage that the term $I(\lambda)$ is canceled out because of the use of the same light source for the RGB camera response and the tristimulus XYZ measured values. The matrix $\mathbf{A}$ is so computed using a least-square linear transform [140]:

$$\mathbf{A} = (\mathbf{RGB}^T \mathbf{RGB})^{-1} \mathbf{RGB}^T \mathbf{XYZ}, \qquad (2.68)$$

where $\mathbf{RGB}$ and $\mathbf{XYZ}$ are, respectively, the matrix form of RGB and XYZ values. Both approaches, as for the methods used for SDR images, need to rely on linear RGB camera responses to avoid the difficulties of treating nonlinear data. Since, DSLR cameras may apply in its rendering pipeline nonlinear transformations (e.g. gamma correction, white balancing, etc.), RAW images are captured.

2.7   Tone Retargeting

An important issue that is still open, is how to map HDR images on SDR displays, and SDR images on HDR displays.

2.7.1   Dynamic Range Compression

Dynamic range compression, typically called *tone mapping*, is the problem of compressing HDR images for displaying on SDR displays or medium (e.g. paper). An operator that compresses the dynamic range of an HDR image is typically called a *tone mapping operator* (TMO), and it is generally defined as

$$f(I) : \mathcal{R}_c^+ \to \mathcal{D}_c, \tag{2.69}$$

where I is an HDR image, c is the number of color channels of I (typically $c = 3$), and $\mathcal{D} \in [0, 2^{n_{\text{bit}}} - 1]$ (typically $n_{\text{bit}} = 8$). However, f typically works on the luminance channel; i.e., $c = 1$, leading to

$$f(I) : \mathcal{R}^+ \to \mathcal{D}, \tag{2.70}$$

where color information is processed in a separate pass; see Section 2.7.3.

Many TMOs have been proposed in the literature in more than 25 years. Although they all have the same main goal, reproduction on a SDR medium, they can be tailored for different subgoals. For example, some operators try to simulate the perception of the HVS and its limits [88, 142, 178, 213, 270]. Other TMOs are designed to create pleasant images that preserve the image local and/or global contrast [74, 86, 230]. Recently, TMOs meant for HDR video backward compression have started to emerge [147, 170, 194].

Despite their subgoals, TMOs are typically classified based on how the processing is handled. Two broad classes exist: point-wise techniques or global operators, and local techniques or local operators.

Point-wise techniques. These TMOs usually apply a constant function, f, to each pixel in the image, and they are based on global statistics of the input luminance channel such as maximum, minimum, average, and geometric average values. In many cases, to be robust to outliers, the maximum and minimum values are respectively computed as the 99th and the 1st percentiles. Geometric average luminance is typically defined as

$$\overline{L}_w = \prod_{i=1}^{N} \big(L_w(\mathbf{x}_i) + \epsilon \big)^{\frac{1}{N}} = \tag{2.71}$$

$$= \exp\left(\frac{1}{N} \sum_{i=1}^{N} \log \big(L_w(\mathbf{x}_i) + \epsilon \big) \right), \tag{2.72}$$

where N is the number of pixels in the image, $\mathbf{x}_i$ are the i-th pixel coordinates, and $\epsilon > 0$ is a small positive value to avoid singularities for luminance values close to or equal to zero.

f can be a general point-wise function, and it is typically a composition of classic range compression functions such as logarithms, exponentials, sigmoids, power functions, etc.

Figure 2.22. An example of photographic tone reproduction operators [230]: On the left side, the simple operator, Equation 2.73. On the right side, the modified version that artistically allows us to burn out bright areas of the image. Note that the bottles are mapped to white to catch viewer attention.

A very popular operator [105], which produces appealing results, is the photographic tone reproduction operator [230] defined as

$$L_d(\mathbf{x}) = \frac{L_s(\mathbf{x})}{L_s(\mathbf{x}) + 1}, \qquad \text{where} \qquad L_s(\mathbf{x}) = a \frac{L_w(\mathbf{x})}{\overline{L_w}}, \tag{2.73}$$

where a is a parameter related to the scene *key value* which typically varies in the range $[0.18, 0.36]$; the lower the value the darker, and the higher the value the brighter. Note that this range is the middle gray range, because the goal is to map key values to middle gray values to take into account the photographic principle of Ansel Adams' Zone System. Moreover, to burn out bright areas in an artistic fashion, Equation 2.73 can be modified as

$$L_d(\mathbf{x}) = \frac{L_s(\mathbf{x})\left(1 + \frac{L_s(\mathbf{x})}{L_{\text{white}}}\right)}{L_s(\mathbf{x}) + 1}, \tag{2.74}$$

where L_{white} is the smallest values that will be mapped to white. Another popular choice is to use a logarithmic function. This is also because a logarithm-based TMO can be linked to the approximation of the Weber–

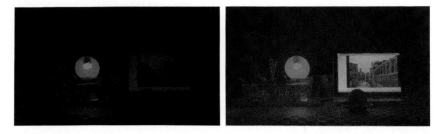

Figure 2.23. An example of TMOs based on logarithmic functions: On the left side, a straightforward mapping of Equation 2.76 is shown. Note that details in dark areas are not well preserved and the overall appearance is dark. On the right side, the adaptive logarithmic mapping, Equation 2.77, is shown. Note that all areas of the image are well reproduced without loss of global appearance.

Fechner law which is defined as

$$k_1 = \frac{B}{\log_e\left(\frac{L}{L_0}\right)}, \qquad \text{or} \qquad B = k_1 \log_e\left(\frac{L}{L_0}\right), \qquad (2.75)$$

where k_1 is a constant factor that can be determined by psychophysical experiments, B is the brightness or the response of the HVS to a luminance stimulus, L, on a background, L_0. A very straightforward approach to map HDR values in the range $[0, 1]$ is to apply a logarithm using as a base the luminance maximum in the image [255], which is equivalent to:

$$L_d(\mathbf{x}) = \frac{\log(L_w + 1)}{\log(L_{w,\max} + 1)}, \qquad (2.76)$$

but this can lead to loss of some high-contrast content and does not provide any user control of the image quality. Therefore, Drago et al. [72] proposed a mapping in which the logarithmic base depends on each pixel's radiance to preserve fine details with a single user parameter. This mapping is defined as

$$L_d(\mathbf{x}) = \frac{L_{d,\max}}{100 \log_{10}(L_{s,\max})} \cdot \frac{\log(L_s(\mathbf{x}) + 1)}{\log\left(2 + 8\left(\frac{L_s(\mathbf{x})}{L_{s,\max}}\right)^{\log_{\frac{1}{2}} b}\right)}, \qquad L_s(\mathbf{x}) = \frac{L_w(\mathbf{x})}{\overline{L_w}},$$

$$(2.77)$$

where $L_{d,\max}$ is the maximum output luminance of the LDR display and $b \in [0.7, 0.9]$ is a control parameter for adjusting brightness.

Other popular approaches in tone mapping are histogram based [150, 154]. In these algorithms, an HDR histogram in the logarithmic domain is first computed using, typically, between 256 and 1024 bins. Then, a mapping function that exploits the HDR histogram knowledge is applied to all pixels.

Although pixel-wise techniques can achieve a very pleasant result and reproduce the overall global contrast, they typically fail to reproduce local and/or fine details.

Local techniques. These techniques apply a spatially varying function, f_s, which takes as input local and global statistics. This preserves the local contrast and details that are typically lost or washed out in point-wise techniques. Practically, the shape of f_s can have shapes similar to the ones used in point-wise techniques, where global statistics are partially or totally replaced by local statistics.

For example, the Reinhard's sigmoid in Equation 2.73 can be made local by replacing the local value in the denominator:

$$L_d(\mathbf{x}) = \frac{sL_w(\mathbf{x})}{s\overline{L}(\mathbf{x}) + 1}, \qquad s = \frac{a}{\overline{L}_w}, \tag{2.78}$$

where $\overline{L}(\mathbf{x})$ is a local average that can be computed using an average or a Gaussian filter; see Appendix A. However, computing statistics locally can

Figure 2.24. An example of local TMOs: On the left side, the TMO defined in Equation 2.78 using a linear filter for computing the local average. Note that a black halo has appeared around the desk lamp. On the right side, the TMO defined in Equation 2.78 using an edge-aware filter; in this case a bilateral filter; see Appendix A. Note that halos are now gone.

introduce artifacts such as halos, which are typically noticed around the edges of very bright areas; see Figure 2.24 (left). This is due to the fact that the result of filtering near the edges can be biased by the contributions of pixels with large differences in intensities from both sides of the edge. This issue can be reduced by employing edge-aware filters such as difference of Gaussians filters [16, 230], multi-scale approaches [13, 85, 157, 211], the bilateral filter and its variants [53, 74], gradient domain editing [86], the weighted least squares optimization [81], total variational structure texture decomposition [256], etc. A popular choice that produces satisfactory results is the bilateral filter [252, 268]; see Appendix A.

Generalized techniques. A novel trend in tone mapping is to propose solutions that are more flexible, general, and robust.

Regarding robustness, the *exposure fusion* [185] is a family of operators that does not require a radiance map as input but all different exposure images. This avoids issues in precisely estimating the CRF. The basic method [185] computes for the i-th exposure a weight map, W_i, which specifies the weight of a pixel in the final image. This can be computed using different metrics such as well-exposedness, color saturation, and edges. Once weights are computed, all exposures are combined together using Laplacian blending defined as

$$\mathbf{L}^l\{I_d\}(\mathbf{x}) = \sum_{i=1}^{n}\mathbf{L}^l\{I_i\}(\mathbf{x})\mathbf{G}^l\{W_i\}(\mathbf{x}), \qquad (2.79)$$

where I_i is the i-th exposure image, I_d is the tone-mapped image, $\mathbf{L}^l$ and $\mathbf{G}^l$ are, respectively, the operator for creating a Laplacian and a Gaussian pyramid [41]. This family of operators generally has two main advantages: it is less prone to color saturation; see Section 2.7.3, and preservation of fine details. However, global contrast in some areas of the image can be lost. An example of exposure fusion is depicted in Figure 2.25.

Figure 2.25. An example of exposure fusion: the first three images from left are SDR exposure images, and the last image is the result of exposure fusion [185].

On the generalization side, Mantiuk and Seidel [179] have proposed a TMO, the generic TMO, which can mimic, with satisfactory accuracy, any TMO by noticing that a TMO can be decomposed into three functions:

$$I_d = MTF(TC(L_w))\left(\frac{I_w}{L_w}\right)^s, \qquad (2.80)$$

where I_w and I_d are, respectively, the color input HDR and SDR images, MTF is the modulation transfer function, TC is a global tone curve applied to each pixel separately, and s is the color saturation. TC is a global curve' that is a four-segment sigmoidal function defined as

$$L_d = TC(L_w) = \begin{cases} 0 & \text{if } L' \leq b - d_l, \\ \frac{1}{2}c\frac{L'-b}{1-a_l(L'-b)} + \frac{1}{2} & \text{if } b - d_l < L' \leq b, \\ \frac{1}{2}c\frac{L'-b}{1+a_h(L'-b)} + \frac{1}{2} & \text{if } b < L' \leq b + d_h, \\ 1 & \text{if } L' > b + d_h, \end{cases} \quad (2.81)$$

where $L' = \log_{10} L_w$, d_l is the lower mid-tone range, d_h is the higher midtone range, b is the brightness adjustment parameter, c is the contrast parameter, and a_l and a_h respectively affect the contrast compression for shadows and highlights. While TC simulates global TMOs, MTF simulates spatially varying aspects of local TMOs. This function specifies, using five basis functions, which spatial frequencies to amplify or compress. A TMO can be

Figure 2.26. An example of combining different TMOs using the hybrid TMO [26] applied to an HDR image: on the left side, the result of Drago et al.'s operator [72]. In the middle, the result of the hybrid TMO [26]. On the right side, the result of Reinhard et al.'s method [230]. Note that the hybrid TMO mixes features from both TMOs according to results of a psychophysical experiment.

simulated by estimating s, and the TC and MTF parameters through the Levenberg–Marquardt fitting procedure. The work done with the Generic TMO has shown how most TMOs have a common ground in terms of image processing, and large differences lie in how TMO parameters are chosen.

Still on the generalization aspect, Mantiuk et al. [175] have proposed another work on TMO generalization on the display side. This operator adaptively adjusts HDR and SDR images given the characteristics of a particular display technology such as SDR monitors, HDR displays, paper, etc. This is achieved by minimizing certain features to be preserved using an HVS-based metric.

Regarding flexibility, researchers have proposed to combine different TMOs using Laplacian blending [41] as in the exposure fusion approach [185]. TMOs can be either selected and weighted based on psychophysical experiments [23] or based on a quality index [305]. These approaches produce appealing images by using the best of each TMO for certain luminance areas; see Figure 2.26. However, they are computationally expensive because many TMOs need to be computed, including their blending weights (which may require running computationally expensive metrics), before Laplacian blending.

2.7.2 Dynamic Range Expansion

Dynamic range expansion, typically called inverse/reverse tone mapping, is the problem of expanding SDR images for displaying on HDR monitors or use in HDR applications. An operator that aims at that is typically called an expansion operator (EO) or inverse/reverse tone mapping operator (iTMO/rTMO); this can be formulated as

$$e(I) : \mathcal{D}_c \to \mathcal{R}_c^+, \tag{2.82}$$

where c is the number of color channels (typically $c = 3$), and $\mathcal{D} \in [0, 2^{n_{bit}} - 1]$ (typically $n_{bit} = 8$). However, e typically is applied only to the luminance channel leading to

$$e(I) : \mathcal{D} \to \mathcal{R}^+. \tag{2.83}$$

The dynamic range expansion problem is an ill-posed problem because there is no data in the overexposed and underexposed areas of an LDR image. However, e is meant to try to recover some information in order to provide an HDR experience with SDR images; see Figure 2.27. As the first step,

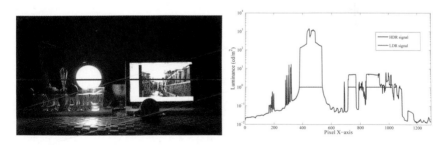

Figure 2.27. The idea behind dynamic range expansion: On the left side, a single exposure image. On the right side, a graph showing the luminance scan-line (in green) from the single exposure image (left). In the graph: the red line shows the full luminance values when capturing a full HDR image; the green line shows the clamped luminance values when capturing an SDR image. Dynamic range expansion techniques provide tools for trying to recover the red profile in the graph starting from the green profile.

most of them require linearization of the input image. This is not an issue, because a lot of SDR content is either encoded with a gamma value of 2.2 (e.g. TV and DVDs) or encoded using sRGB. This information can be easily

retrieved from the file metadata; e.g. the EXIF file in the case of JPEG.
In the worst case, when the CRF is unknown or there is no metadata, the
CRF or gamma encoding can be estimated. For example, gamma encoding
can be estimated using bispectral analysis [82]. Note that the CRF can
be roughly estimated from a single image by exploiting information at the
edges [158, 159]. Another common step is to gently filter the image with
an edge-aware filter (e.g. the bilateral filter) before expanding the dynamic
range. This reduces contouring effects and noise in the image, increasing
the quality of the final result.

Point-wise techniques. These apply the same expansion function to
the whole image. Akyüz et al. [6], through a series of perceptual experiments,
showed that a simple linear scaling of a well-exposed high-quality image
without compression can provide a satisfactory HDR experience. The scaling
is defined as

$$L_w = k\left(\frac{L_d - L_{d,min}}{L_{d,max} - L_{d, min}}\right)^{\gamma}, \tag{2.84}$$

where $\gamma = 1$, and k is the maximum luminance of the target HDR display.
Masia et al. [181] showed that when the image is overexposed, γ has to
be adjusted to the content. Therefore, through a series of psychophysical
experiments, they proposed a content-dependent γ as

$$\gamma(k) = ak + b \qquad k = \frac{\log L_{H, d} - \log L_{d,min}}{\log L_{d, max} - \log L_{d, max}}, \tag{2.85}$$

where $a = 10.44$ and $b = -6.282$ are the data-fitting results of their
psychophysical experiments. Note that γ can be negative for some $L_{H, d}$
values leading to negative images.

Classification techniques. These classify the image into different
large areas, and a different expansion function is applied for each area.
For example, Meylan et al. [186] proposed to apply simple automatic
thresholding for classifying highlights and light sources from diffuse surfaces.
Then, two different linear functions are applied to these two areas. However,
this fully automatic method can produce unsatisfactory results in some
cases when thresholding fails. For example, the overexposed sun and a
white wall will both be mapped to the maximum expansion value leading
to an unexpected perception. Didyk et al. [70] improved on this work using
machine learning and a more sophisticated expansion operator based on a
histogram equalization mechanism that exploits image gradients.

Non-Local techniques. These methods typically generate an *expand
map*, which is a smooth and possibly edge-aware field representing areas of
the image that are likely to be expanded; see Figure 2.28. The expand map
can be computed in many different ways:

- Sampling and Filtering: The luminance channel is sampled using
  importance sampling techniques [62] to generate samples where light

Figure 2.28. An example of an expand map computed using Banterle et al.'s method [25]. On the left side, an input LDR image. On the right side, the resulting expand map.

sources can be. Then, density estimation is performed [24] and an edge-aware filter is applied to this result [25].

- Thresholding and Filtering: The luminance channel is thresholded using a typical saturation value in the SDR image (between 230 and 250), and then filtered using an edge-stopping or edge-aware filter [146, 231].

- Filtering: The luminance channel is filtered using an edge-aware filter [120].

Once an expand map is computed, it is used to drive expansion, which is achieved using point-wise techniques such as linear expansion [231], power functions [120], inverting a tone curve [24], etc.

The two main advantages of local operators is the ability to reconstruct lost luminance profiles, because the expansion varies across the image, and to attenuate quantization or compression artifacts. This is because underexposed images are typically left untouched.

Synthesis from other images. A different approach to dynamic range expansion is HDR hallucination [280]. This technique exploits texture synthesis [31] to fill missing information in overexposed and underexposed areas. Moreover, users can inpaint details from the same SDR image or other images using a stroke-based interface using Poisson image editing [216]. This method tends to have high-quality results, but it requires heavy user inputs.

2.7.3 Color Retargeting

Given an HDR input image specified in the linear RGB color space (RGB_w), the tone retargeting first maps its luminance values, L_d, into a smaller luminance dynamic range, L_d, typical of a display system, as described in Section 2.7.1. A common approach to adjust saturation is to post-process the tone-retargeted image by preserving the color ratio [243]:

$$RGB_d = \frac{RGB_w}{L_w}L_d. \tag{2.86}$$

Strong contrast retargeting may result in oversaturated images; see Figure 2.29. To solve this issue, Schlick et al. [243] proposed an *ad hoc* approach based on the power law:

$$RBG_d = \left(\frac{RBG_w}{L_w}\right)^s L_d, \tag{2.87}$$

where s is a saturation parameter; $s > 1$ means an increase in saturation,

(a)   (b)   (c)

Figure 2.29. An example of different saturation values, s, after tone mapping: (a) $s = 0.5$, (b) $s = 1.0$, and (c) $s = 1.5$.

while $s < 1$ means a decrease in saturation. The major drawback of this solution is the introduction of undesirable luminance and hue shifts. Mantiuk et al. [174] proposed an alternative approach that interpolates between chromatic and corresponding achromatic colors. This is defined as

$$RBG_d = \left(\left(\frac{RBG_w}{L_w} - 1.0\right)s + 1.0\right)L_d. \tag{2.88}$$

This approach reduces luminance shift, but it may still produce hue shift [174]. The parameter $s \in [0, 1]$ can be manually set with the disadvantage that it depends on both image and tone reproduction operators [222]. s can be estimated for both approaches, Equations 2.87 and 2.88, if the luminance tone curve is given. However, the luminance shift introduced when applying Equation 2.87 and the hue shift when applying both Equations 2.87 and 2.88 cannot be fully removed. In particular, the relationship between saturation s and contrast c has been derived by fitting the statistical results of a psychophysical experiment with a sigmoid function:

$$s(c) = \frac{(1 + k_1)c_2^k}{1 + k_1 c_2^k},$$ (2.89)

where k_1 and k_2 are derived by the best least-square fit; for Equation 2.87 we have $k_1 = 1.6774$ and $k_2 = 0.9925$ and for Equation 2.88 we have $k_1 = 2.3882$ and $k_2 = 0.8552$. Notice that if contrast is not modified, the function does not change the color saturation; i.e., the case $s = 1$ and $c = 1$. If the image has no contrast, $c = 0$, it removes all color information, i.e., $s = 0$.

Color Correction for Tone Reproduction. Starting from the original HDR image (I_o), the original hue and saturation can be inferred into a color-distorted tone-mapped image [222] (I_t). The main goal is to match the I_o color appearance in terms of hue and saturation, while preserving luminance values from the tone-mapped image I_t. The idea is to use a perceptually uniform color space, which minimizes the distortion of the appearance attributes of the HDR image (see Section 2.3 and Section 2.4.1). Therefore, the IPT color space has been adopted for image representation because it can represent hue uniformity. Then, the image is converted into a cylindrical color space, CIE $L^*C^*h^*$ [248], which allows the direct manipulation of its appearance attributes: lightness I, chroma C, and hue h:

$$I = I_{IPT}$$ (2.90)

$$h = \tan^{-1}\left(\frac{P}{T}\right),$$ (2.91)

$$C = \sqrt{P^2 + T^2}.$$ (2.92)

Two aims must be met. First, the lightness of the tone-mapped image needs to be preserved. To achieve this, the lightness channel I is not further processed. Second, the hue distortion of the original HDR image needs to be minimized while matching the saturation of the tone-mapped image to the one of the original HDR image. Since the hue of the tone-mapped image, h_t, may be distorted by gamut clipping during tone-mapping, it is set to h_o, the hue of the original HDR image. Saturation S is computed through a formula that mimics the human perception more closely [168]:

$$S = \frac{C}{\sqrt{C^2 + I^2}}.$$ (2.93)

The tone-mapping step compresses luminance in a nonlinear manner without modifying the chromatic information. Since saturation depends on lightness, modifying luminance (or lightness) will surely change the saturation of the tone-mapped image leading to an oversaturated effect [222]. The nonlinear luminance mapping during tone-mapping is the cause of an increment of relative luminance for several pixels of the tone-mapped image when compared with their surrounding pixels [222]. To deal with this mismatch, the chroma C_t of the tone-mapped image is scaled to approximately what it would be if the original HDR image had been tone-mapped in the CIE $L^*C^*h^*$ color space:

$$C_t' = C_t \frac{I_o}{I_t}. \tag{2.94}$$

Then, based on Equation 2.93, the ratio r between the saturation of the original and tone-mapped images is computed:

$$r = \frac{S(C_o, I_o)}{S(C_t', I_t)}. \tag{2.95}$$

This ratio is then applied to chroma C_t' to find the chroma-corrected C_c for the tone-mapped image:

$$C_c = rC_t' = r\frac{I_o}{I_t}C_t. \tag{2.96}$$

The hue of the tone-mapped image is reset by copying values from the HDR images ($h_c = h_o$). The chroma-corrected C_c is combined with the lightness channel of the tone-mapped image $I_c = I_t$ to produce the final corrected result, which can then be converted back to RGB.

2.8   Gamut Mapping vs. Tone Mapping

Gamut mapping and tone mapping have often been classified under the same category of techniques that share the goal of mapping a large set of colors into a smaller one. However, they differ in many ways. First, tone mapping is typically used to prepare HDR images for display on SDR display devices, when the dynamic range of the input image is vastly higher than the dynamic range of the display device. Second, the goal of tone mapping is to compress luminance values, while the gamut mapping goal is to preserve, as much as possible, the perceptual attributes of lightness, hue, and chroma. In more detail, these differences can be identified in the following aspects:

- **Dynamic Range**: Considering just the luminance range, gamut mapping typically works with a limited dynamic range where the luminance is in between the range $[0, 100]$ cd/m^2, while tone mapping deals with a vast luminance dynamic range that covers several orders of magnitude of the range available in the real world.

- **Color Gamut**: As in the luminance range, large discrepancies are also within the color gamuts that tone mapping deals with. Large differences exist between the color gamut of the input HDR image and the color gamut of a typical SDR display such as sRGB. Gamut mapping works with much lower differences in the gamut boundaries between the source and the destination gamuts.

- **Rendering Goal**: For tone mappings the main goal is to compress luminance while preserving contrast and details of the input image, color fidelity reproduction is not its major goal. Post-processing steps are adopted to reaugment the color information. As consequence of this, the tone-mapped image may still contain out-of-gamut colors, which are clipped to gamut boundaries in an uncontrolled manner. However, in gamut mapping the aim is to preserve the perceptual attributes through the minimization of the hue shift. Color fidelity reproduction is its major goal, while limiting out-of-gamut colors and preserving color contrast.

2.9 Gamut Mapping Approach for HDR Content

Tone mapping itself limits compression to luminance values only. This may generate changes in the chroma and hue of the original HDR image causing changes in the color appearance of the tone-mapped image. Typically, these issues are managed with post-processing techniques as shown in Section 2.7.3. However, these techniques do not take into account the destination gamut where the tone-mapped image is finally displayed [276]. As a consequence of this we may either have pixels of the tone-mapped image left out from the display gamut, or if these pixels are mapped into the display gamut, strong hue shift and luminance distortion are generated.

To solve these issues, Šikudová et al. [276] have proposed a gamut mapping framework for HDR content. In this framework, tone mapping is integrated with gamut mapping, retargeting the colors after dynamic range compression, and ensuring that the image fits within the display gamut. The main advantage over the existing color retargeting techniques, for HDR content, is that the hue shift and luminance distortion introduced by their framework is limited when compared with existing techniques.

Figure 2.30 shows the framework and its four main steps. First, the input HDR image is tone mapped, where the tone mapping consists of compressing Y luminance values. After that, the input HDR image is converted to the Yxy color space. Once tone mapped, the luminance channel Y is substituted in the original XYZ values of the input image. Y is normalized so the maximum value is 100 cd/m^2. As shown in Section 2.3.4, the choice of a working color space, in the context of gamut mapping, is essential for reducing hue and chroma shift. Several color spaces exist such as CIE $L^*a^*b^*$, CIE $L^*C^*h^*$, and IPT. In this work, the authors chose the CIE $L^*C^*h^*$ as a working color space. At this point, gamut boundaries need to be computed for both source (input HDR image) and destination (display) gamuts. This helps identify if pixels of the source gamut are either within or out of the destination gamut, and compute the mapping directions needed for the gamut mapping function. Since the main goal

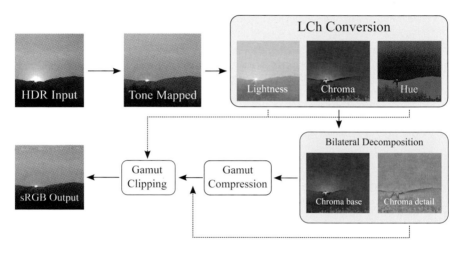

Figure 2.30. The HDR gamut mapping framework proposed by Šikudová et al. [276].

is to reduce, as much as possible, the hue shift, the h component is left untouched while compression is performed only on chroma C. To achieve this, bilateral filtering (see Appendix A) is used to preserve edges in the C channel, while the compression is performed on the base layer. Once the chroma C compression is performed, the details are added back, and a smooth clipping is applied. The gamut clipping is needed because separately processing lightness and chroma cannot guarantee that all pixels are within the destination gamut. Figure 2.31 shows this problem. Tone mapping guarantees that the maximum value of lightness is 100 cd/m^2; however, pixels are still out of the destination gamut in their chroma direction. Depending on how chroma C is compressed, we may still have a few pixels out of the destination gamut. Finally, the clipped corrected version of the CIE $L^*C^*h^*$ image is converted back to the sRGB color space. The C compression proposed in [276] follows two distinct approaches. The first one, named the hue-specific method, can maximize the use of the

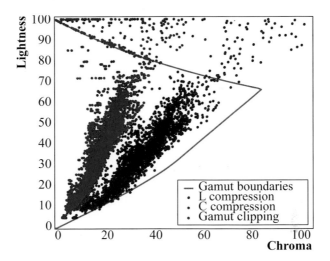

Figure 2.31. Tone mapping does not guarantee that all the pixels of the tone-mapped image are within the destination gamut where the image will be visualized. Pixels may be within the destination gamut only for the lightness channel L^*; however, their chroma channel may still be out of gamut. Image courtesy and copyright of Tania Pouli.

available gamut while fully compressing the chroma. However, this requires a high computational complexity. A simpler and quality-efficient chroma compression method, the global approach, has been proposed, which have a lower computational complexity. This method relies on the interesting observation that the dynamic range of the chroma channel is not extremely high, when compared to the dynamic range of the destination gamut. This suggests that linear compression may be enough [276]. As indicated above, the chroma compression is applied only to the base layer of the chroma channel, $C_{\mathrm{base},h}$, as follows:

$$C'_{\mathrm{base},h} = C_{\mathrm{base},h} \min_{h \in [0°,359°]} \left[\frac{\mathrm{Cusp}_{d,h}}{\mathrm{Cusp}_{s,h}} \right], \qquad (2.97)$$

where $\mathrm{Cusp}_{s,h}$ and $\mathrm{Cusp}_{d,h}$ are, respectively, the maximum chroma values for the source and destination gamuts at a specific hue angle h. In other words, the compression is performed at each hue angle h with an incremental step of 1 degree. Note that only the source gamut is compressed and only when $\min \left[\mathrm{Cusp}_{d,h}/\mathrm{Cusp}_{s,h} \right] < 1$. The chroma compression not only avoids hue shift when compared to clipping techniques, but it also reduces the oversaturated appearance typical of the tone-mapping step. Using the full chroma range of the source gamut may produce extremely compressed chroma results. This is typically due to the use of a few outlier pixels with extremely high chroma values. This problem can be avoided by using a percentile approach that eliminates these outlier pixels [276]. The authors have also proposed a lightness compression strategy, integrated within the proposed framework. In this case, the lightness component L^* is also filtered with a bilateral filter (see Appendix A) and the compression is performed on its base layer, while the details are added back after the compression step.

2.10   Color Appearance for HDR Applications

In the case of HDR content, large discrepancies between viewing conditions of the two environments, acquisition and visualization, exist when compared to SDR content. To fully convey the captured HDR scene appearance when visualizing it on a SDR display, we need to take into account the color appearance attributes and faithfully reproduce them under the viewing conditions where the scene is visualized. To address this issue, we have two options; either making use of existing CAMs before applying tone mapping [7], or extending an existing CAM to treat HDR images [152]. The first type of approach is depicted in Figure 2.32, and it takes as

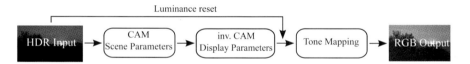

Figure 2.32. The framework proposed by Akyuz et al. [7] for applying CAMs before applying tone mapping.

input a calibrated (absolute values) HDR image in the linear RGB color space [7]. Calibrated data are obtained by multiplying the original HDR input with a scale factor that maps the HDR content into absolute units. This scale factor is computed as a ratio between the luminance measured of uniform gray patches placed into the scene and the luminance values of these patches in the digital image. In case the scale factor cannot be measured, it can be estimated using some heuristics as described in [7]. As shown in Figure 2.32, the input HDR image is given as input to an existing CAM, i.e., CIECAM02. Once, the color appearance attributes are adapted to the new viewing conditions of the SDR display, the image is tone mapped using an existing TMO. The degree of adaptation value D is the one used in CIECAM02. Using Hermite interpolation, the value of D is smoothly interpolated between the upper and lower 10% of a luminance threshold L_t that indicates the presence of a light source. This corresponds to the range of luminance values $L_t \pm 0.1(L_{\max} - L_{\min})$, and the value of L_t is computed as

$$L_t = L_{\min} + (0.6 + 0.4(1 - k))(L_{\max} - L_{\min}). \qquad (2.98)$$

The appearance correlates are computed applying the forward model of the CIECAM02 to the tristimulus XYZ values of the HDR input image. Afterward, these are converted back to the tristimulus XYZ values, and adapted to the viewing conditions of the visualization environment (SDR display), using the inverse model of the CIECAM02. Then the new tristimulus YXZ values are converted to Yxy, where the Y values are substituted with the original values of the luminance Y of the input HDR image. This separates the color appearance step from the tone mapping. Finally, the original luminance values Y are tone mapped, and the new Yxy values are converted back to the RGB color space ready to be visualized; i.e., $sRGB$. The method is independent of any specific TMO and color appearance model [7].

Following the second type of approach, iCAM has been extended to work with HDR images [152], and its framework is depicted in Figure 2.33. As

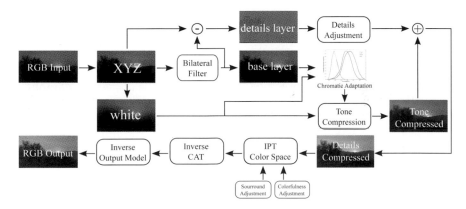

Figure 2.33. iCAM06 pipeline, an extension of iCAM for processing HDR images.

the first step, the input HDR image is converted into a device-independent color space, i.e., XYZ. Then it is decomposed into two scale layers, base and details, using bilateral filtering (see Appendix A). This decomposition is well motivated by two widely accepted assumptions. First, the image can be seen as a product of the reflectance (details layer) and the illumination (base layer); and the HVS is more sensitive to the reflectance than the illumination. Secondly, local contrast relates to the reflectance image, and the HVS responds mostly to local contrast. Therefore, due to the insensitivity to global luminance contrast by the HVS, it is appropriate to process the base layer while retaining the local information such as details.

iCAM06 includes many traditional steps of iCAM, but it also introduces some modifications to process HDR images. The forward model consists of the following steps: the Chromatic Adaptation Section 2.4.2, Post-adaptation Signal Compression Section 2.4.3, and Color Decorrelation Section 2.4.4. At this stage, the inverse model of the color appereance is applied to simulate the viewing conditions of the environment in which the image is visualized.

Forward Model:

- **Chromatic Adaptation**: The chromatic adaptation used by iCAM06 is the same chromatic adaptation used in iCAM. This is the chromatic model adopted by CIECAM02. A factor of 0.3 is applied to the computation of the adaptation factor D, Equation 2.24, for reducing the desaturation effect for the HDR rendering [152]:

$$D = 0.3F\left(1.0 - \frac{1.0}{3.6}e^{-\left(\frac{L_a + 42}{92}\right)}\right). \qquad (2.99)$$

- **Tone Compression**: The tone compression is a mix of rod and cone responses. The cone response prediction is the CIECAM02 post-adaptation mechanism. The cone responses are then compressed similarl to Equation 2.28 used in CIECAM02, where the constant power value 0.73 now is a user-controllable parameter p that modifies the contrast of the output image. Acceptable values are in the range $[0.6, 0.85]$. However, a value of 0.75 has been empirically set [152]. In Equation 2.28, 100 is equivalent to the luminance of the reference white Y_w; but in iCAM06, it is equivalent to the luminance of the local adapted white image. The F_L factor in iCAM06 is a spatially varying function derived from the low-pass adaptation image at each pixel location, while previously this factor was not a spatially varying function.

The rod responses are adapted from the ones used in the Hunt model [119]:

$$A_s = 3.05B_s \frac{400\left(F_{LS}\frac{S}{S_w}\right)^p}{27.13 + \left(F_{LS}\frac{S}{S_w}\right)^p} + 0.3, \qquad (2.100)$$

$$F_{LS} = 3800j^2(5L_A) + 0.2(1 - j^2)^4(5L_A)^{1/6}, \qquad (2.101)$$

$$j = \frac{10^{-5}}{(5L_A) + 10^{-5}}, \qquad (2.102)$$

$$B_s = \frac{0.5}{1.0 + 0.3[(5L_A)(S/S_w)]^{0.3}} + \frac{0.5}{1.0 + 25L_A}, \qquad (2.103)$$

where S is the luminance pixel in the chromatic adapted image, and S_w is the reference white for the image S. The final tone compression response is the sum of the cone and rod responses.

- **Color Decorrelation**: The color decorrelation step is the conversion of the tone-mapped image to the IPT color space and it is done as in the iCAM model Section 2.4.6. The tone-mapped image is the tone compressed responses where the details are afterward added back. To predict the Stevens effect, i.e., an increase of luminance-level results as an increase of local perceived contrast, an adjustment of the detail layer (*details*) is needed and it is performed as follows:

$$details_a = details^{(F_L+0.8)^{0.25}}. \tag{2.104}$$

The IPT components are also enhanced to predict various color appearance phenomena. The phenomenon where an increase in luminance-level results in an increase in perceived colorfulness, also called the Hunt effect, is predicted enhancing the P and T components as follows [152]:

$$P_a = P\left[(F_L + 1.0)^{0.2}\left(\frac{1.29C^2 - 0.27C + 0.42}{C^2 - 0.31C + 0.42}\right)\right], \tag{2.105}$$

$$T_a = T\left[(F_L + 1.0)^{0.2}\left(\frac{1.29C^2 - 0.27C + 0.42}{C^2 - 0.31C + 0.42}\right)\right], \tag{2.106}$$

where C is the chroma appearance correlate calculated as in Equation 2.55. The component I is enhanced to predict the effect that the perceived contrast image increases when the image surround changes from dark to dim to light [152]:

$$I_a = I^\gamma, \tag{2.107}$$

where γ is 1.5 for dark, 1.25 for dim, and 1.0 for average surround.

Inverse Model:

After the IPT components are enhanced to predict various color appearance phenomena, the image needs to be converted for display on an output device. Therefore, the IPT image is converted into the XYZ tristimulus values. An inverted chromatic adaptation step, i.e., the inverse of the CIECAM02 CAT model, to adapt XYZ values to the white point of the display media, XYZ_c, is performed. Then, the chromatic adapted XYZ_c values are converted to the RGB device-dependent color space of the display. Extreme dark and bright values can be removed using a clipping of the 1st and 99th percentile [152]. Finally, gamma correction, to account for the nonlinearity of the display, and scaling between 0 and 255 are required for the final visualization of the image.

2.10.1   High-Luminance Color Appearance Model

Similar psychophysical experiments used to derive color appearance phenomena, under limited luminance dynamic range $[0, 100]$ cd/m^2, can also be used for a higher luminance dynamic range through the use of an HDR display. These psychophysical data can be used to derive a color appearance model for high luminance dynamic range content [141].

The forward model steps are defined as follows:

- **Chromatic Adaptation**: The $CAT02$ chromatic adaptation model used in CIECAM02 is adopted.

- **Cone Response**: After the chromatic adaptation, the tristimulus values are converted to LMS cone space as in CIECAM02 and described in Section 2.4.3. Then, the absolute responses are computed following the Michaelis–Menten equation [141] where the σ factor is substituted with the absolute level of adaptation L_a (average luminance computed at 10 degrees of field of view):

$$L' = \frac{L^{n_c}}{L^{n_c} + L_a^{n_c}}, \qquad (2.108)$$

  the same equation is applied to the other components, M and S. The value of the parameter $n_c = 0.75$ is derived from psychophysical data.

- **Color Decorrelation** - The achromatic signal A is computed using a linear combination of the absolute cone responses, and it is believed that weighting between the three cone responses is $40 : 20 : 1$ [141]:

$$A = (40L' + 20M' + S')/61, \qquad (2.109)$$

  while the opponent signals a and b are computed as in classical $CAMs'$ as CIECAM02 Equation 2.35.

- **Color Appearance Attributes** - The computation of color appearance attributes share several aspects with the way they are computed in CIECAM97 and CIECAM02. Lightness has been roughly defined as the ratio between the achromatic signal A and the achromatic signal for the reference white A_w. From experimental data, the plot of the ratio versus the perceived lightness follows an inverse of the sigmoid function [141]. The goal is to have a linear map of the predicted lightness to the perceived lightness. This is obtained through the application of an hyperbolic function as (lightness J'):

$$J' = g\left(\frac{A}{A_w}\right), \qquad (2.110)$$

where g is defined as

$$g(x) = \left[\frac{-(x - \beta_j)\sigma_j^n}{x - \beta_j - \alpha_j} \right]^{\frac{1}{n_j}}. \qquad (2.111)$$

The values of the parameters are derived from experimental data, yielding $\alpha_j = 0.89$, $\beta_j = 0.24$, $\sigma_j = 0.65$, and $n_j = 3.65$. J' may have negative values. This happens when dark colors cannot be distinguished by the observer so it has to be set to 0 [141]. The perception of lightness is strongly dependent on the media where it is observed. To account for it and to improve the lightness prediction, the lightness J' is modified as [141]

$$J = 100 \Big[E(J' - 1.0) + 1.0 \Big], \qquad (2.112)$$

where E is a parameter that defines the particular medium. Brightness Q is derived as

$$Q = J L_w^{n_q}, \qquad (2.113)$$

where parameter $n_q = 0.1308$ is derived from experimental data [141]. The chroma C is typically computed as a magnitude of the opponent signals:

$$C = \alpha_k \left(\sqrt{a^2 + b^2} \right)^{n_k}, \qquad (2.114)$$

where $n_k = 0.62$ and $\alpha_k = 456.5$ are derived from the experimental data. Colorfulness M has been found to be linear in the logarithm of the reference white luminance L_w [141]:

$$M = C \Big(\alpha_m \log_{10} L_w + \beta_m \Big), \qquad (2.115)$$

where $\alpha_m = 0.11$ and $\beta_m = 0.61$ are derived from the experimental data. Saturation and hue angle are computed as in CIECAM02 Equation 2.45 and Equation 2.36, respectively.

2.11 Summary

The final research goal, in the development of a multimedia system, has always been to convey to the end user the truly real-life experience of the HVS. This consists of simulating various aspects of the HVS such as adaptation processes, color appearance phenomena, etc., as well as to overcome the technology limitations of the current acquisition and visualization devices.

However, no matter how good the proposed solution is in avoiding device technology limitations, we will always have a constraint in the maximum quality that we will be able to reach. A typical example is the gamut mapping problem, described in this chapter. This problem consists of compressing bigger gamut (source) into a smaller gamut (destination) of colors. This implies that the source color will never be mapped exactly to the same color in the destination gamut, but it will be as close as possible. However, solutions that may limit this constraint can play on remapping the entire source gamut in the destination gamut reducing as much as possible color appearance changes. An interesting future trend will be reducing possible device technology limitations, while being able to truly reproduce the color appearance phenomena that we experience daily. Concerning the HDR imaging pipeline, recently, great interest has been demonstrated by the entertainment industry. In particular, many activities in standardization communities are actively proposing new standards for HDR backward compatibility with existing formats, e.g., JPEG XT [11, 12, 144, 232]. This will open an exciting future, where acquisition devices can directly acquire HDR content and truly store it, which will be visualized as tone-mapped content when visualized on SDR display devices.

3

Color2Gray

3.1 Introduction

In the context of imaging systems, we may often meet some constraints that limit the way that we typically transmit, store, and visualize the image content. In the case of black-and-white printing systems; a color image has to be converted into a meaningful gray-scale image. Another example is the photorealistic rendering on black-and-white media [141]; this system may reproduce the output color image, but it needs to be converted to a gray-scale image for visualization purposes. The major aim of this conversion is to convey the same information contained in the original color image. This means preserving the appearance of the original color image while maintaining salient information such as edges, details, contrast, etc. The color2gray conversion problem can be seen as the classical dimensionality reduction problem, where an n-dimensional space is converted into an m-dimensional space, where $n > m$. Unfortunately, traditional techniques, e.g., using the luminance channel Y as, gray-scale image, are often failing in preserving the appearance of the original color image, i.e., important features are lost. We will formalize the problem of color2gray conversion as the typical dimensionality reduction problem. Then we will introduce some naïve approaches, followed by more sophisticated techniques with the aim of preserving specific features of the original color image. Finally, the inverse problem, colorization, will be briefly introduced.

3.2 Dimensionality Reduction Problem

Starting with an n-dimensional space as input $I = (I_1, I_2, \ldots, I_n)^T$, we want to find an m-dimensional space, $G = (G_1, G_2, \ldots, G_m)^T$, where $n > m$. G has to retain, based on some constraints, as much as possible the original content of I [122]. For simplicity, we represent the problem in a matrix form. Considering that we have p observations and each of these has n

dimensions, the problem can be expressed with the following matrix form:

$$I = \mathbf{M}G, \tag{3.1}$$

where G is a $m \times p$ matrix, I is a $n \times p$ matrix, and $\mathbf{M}$ is a $n \times m$ matrix. This is a classical linear relationship in a matrix form, where the new m-dimensional space is expressed as a linear combination of the input n-dimensional space. The matrix $\mathbf{M}$ is also called the linear transformation weights matrix [122]. The problem can be solved with classical linear models. However, if the relationship between I and G is strongly nonlinear, nonlinear models are required to be used, and the problem of Equation 3.1 needs to be formulated as follows:

$$(I_1, I_2, \ldots, I_n)^T = f((G_1, G_2, \ldots, G_m)^T), \tag{3.2}$$

where f is a nonlinear mapping function.

3.2.1   Color2Gray

Color2gray is a typical example of the dimensionality reduction problem. In this case, the matrix I is a three-dimensional input image with color coordinates R, G, and B, and the goal is to produce a 1-dimensional image as output, the reduced dimensional space G. The difficulty of the problem lies in the fact that we want to reduce the dimensionality of a color image into a 1-dimensional image and to convey the same original appearance to the observer. Figure 3.1 shows how this problem is complex and difficult. In this case, the color input image is converted to a 1-dimensional space making use of the ITU Rec.709 conversion equation from RGB to luminance Y:

$$Y = 0.2126R + 0.7152G + 0.0722B. \tag{3.3}$$

Relevant information that is clearly visible in the color input image Figure 3.1 (a), can be lost; see Figure 3.1 (b). In the case of isoluminant colors, colors that have the same luminance information Y, lose all information in the chromaticity dimensions when converted to Y; see Figure 3.1 (c).

The major aim of a color2gray mapping is to reproduce a perceptually accurate gray-scale image. This means that the mapping function needs to emulate both global and local behavior [251]. In particular, this consists of preserving the dynamic range, the average luminance of the input color image, and the local contrast. Moreover, the gray values need to be ordered retaining the original appearance of the input color image and spatial details need to be preserved as much as possible making the possible difference imperceptible [251]. Following these principles introduced by Smith et al. [251], we can impose one or more constraints for making the problem treatable, and at the same time, to reproduce as much as possible the visual appearance of the input color image into the gray-scale image:

Figure 3.1. Relevant information is lost when the RGB input color image is converted to 1-dimensional space. In this case, the input RGB color image (a–c) is converted to luminance Y making use of the standard formula of the ITU Rec.709 (b–d). In the case of isoluminant colors (c), the information loss is severe (d). Input images (c) courtesy and copyright of Laszlo Neumann.

- **Preserve Contrasts**: [228] The distance between any color pair in the input image is proportional to the distance between their gray values. This can also be achieved preserving the dynamic range of the input color image [103]. This concept can also be linked to the contrast magnitude, where the gray-scale contrasts should visibly reflect the magnitude of the color contrasts [103].

- **Luminance Consistency**: [228] This constraints allow avoiding luminance reversal, also called image polarity effects (see Figure 3.2) when the color image is converted to gray-scale. This can be achieved through preservation of the luminance ordering. Grundland et al. [103] pointed out that when pixels in the input image, with increasing luminance, share the same hue and saturation, they need to be mapped to pixels with increasing gray levels in the gray-scale image. We may also make this constraint even stronger by imposing that a gray level in the input color image needs to be mapped into the same gray level in the gray-scale image [103].

- **Mapping Consistency**: The same colors should be mapped to the same gray-scale value [141]. This is also called global consistency [103]. This is a global property of the mapping where the ordering of the gray levels reflects the global ordering relation in the color image [103].

- **Features Preservation**: Features, such as details and edges, in the input color image should remain visible in the gray-scale image [141].

- **Preserve Ordering**: The gray-scale image should respect the ordering of colors of the input image [141]. This can also be seen as hue and saturation ordering [103].

- **Lightness Stimuli**: The color input and the gray-scale images should have as similar as possible lightness stimuli [141].

- **Contrast Polarity**: The aim is to retain the polarity of luminance change in the color contrast of the input image [103].

- **Continuous Mapping**: To avoid false contour artifacts in homogeneous color regions, the mapping function needs to be a continuous function [103].

3.3   Principal Component Analysis (PCA)

Principal component analysis (PCA) is a linear technique that is commonly used to solve the linear problem specified in Equation 3.1. PCA identifies an orthogonal linear combination of the input variables with the largest variance (principal components). In other words, it identifies the principal directions where the input variables vary. This means that the first principal component, which is the first linear combination with the largest variance, is the variable containing the most information. Principal components are

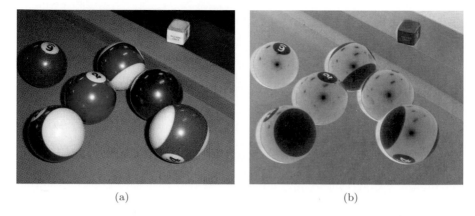

(a)   (b)

Figure 3.2. An example of luminance reversal artifacts generated during the color2gray mapping. The luminance values are swapped such that the highest and lowest luminance values in the input image (a) are mapped to black (lowest) and white (highest), respectively, in the gray-scale image (b).

computed by eigenvalues and eigenvectors of the image covariance. Since the variance depends on the scale of the variables, we need to normalize the variable scale to have comparable measurements. This is achieved by centering the image covariance to the image mean. Starting from an input image I with p observations (or pixels) and n dimensionality (or color channels), we first compute the input image covariance C, an $n \times n$ matrix, as

- Centering the image, I, to its mean:

$$U = I - I_{\text{mean}}. \tag{3.4}$$

- Image covariance computation:

$$C = U^T U / (n - 1). \tag{3.5}$$

To compute eigenvalues and eigenvectors, the singular value decomposition (SVD) method can be used. This decomposes the input image covariance matrix as

$$C = MDM^T, \tag{3.6}$$

where M is an orthogonal $n \times n$ matrix whose columns are eigenvectors of C, and D is a diagonal matrix whose diagonal elements are the eigenvalues of C. Sorting the variances in decreasing order, the output image G with reduced dimensionality m is obtained as the first principal component:

$$G = M^T U. \tag{3.7}$$

A typical drawback of PCA is the luminance reversal artifacts as shown in Figure 3.2. This is due to the fact that eigenvectors can be negative. To avoid this drawback, we need to check that the ordering of the input image luminance matches the ordering of the eigenvectors. If this is not respected, the sign of each eigenvector needs to be changed.

3.3.1   Kernel PCA

The major drawback of PCA is that other principal components may contain small but significant features that should not be discharged. This is shown in Figure 3.3. Kernel PCA can derive features variation more efficiently

RGB Color input image   Zoomed area

PCA gray-scale image   Zoomed area

Figure 3.3. Typical problem of PCA in color2gray mapping is the loss of features that are contained in other principal components. Loss of color variations are not captured by the PCA as shown in the zoomed areas (red).

than PCA [307]. In this case, the kernel matrix is introduced and used instead of the covariance input image matrix C:

$$K_{i,j} = K(y_i, y_j), \qquad (3.8)$$

where $K(y_i, y_j)$ is the kernel function of (y_i, y_j), and y_i and y_j are n-dimensional row vectors of the input image. Different kernel functions can be used; examples are given in Zhang et al.'s work [307]:

- **Polynomial**: Example of quadratic polynomial is given:

$$K_{i,j} = \big(a(y_i \cdot y_j) + b\big)^2. \qquad (3.9)$$

- **Radial Basis Function (RBF)**:

$$K_{i,j} = \exp\left(-\frac{||y_i - y_j||^2}{2\sigma^2}\right). \qquad (3.10)$$

- **Sigmoid**: Also called the hyperbolic tangent function:

$$K_{i,j} = \tanh\big(\alpha(y_i \cdot y_j) + b\big). \qquad (3.11)$$

- **Cosine**:

$$K_{i,j} = \frac{\big(a(y_i \cdot y_i) + b\big)^2}{\big(a(y_j \cdot y_j) + b\big)\big(a(y_i \cdot y_i) + b\big)}. \qquad (3.12)$$

The parameters a, b, α, and σ are set according to experimental tests [307].

3.4   Image Gradient

The image gradient is one of the most popular tools used in image processing to extract local information. It measures a color directional change in an image. Considering a 2D image I as a function, its gradient is typically computed as its derivatives in the x and y directions [98]:

$$\nabla I = \frac{\partial I}{\partial x}\hat{x} + \frac{\partial I}{\partial y}\hat{y}, \qquad (3.13)$$

where $\hat{x}$ and $\hat{y}$ are standard unit vectors. Since the image is a discrete function, the derivatives can be approximated convolving an image with a filter kernel h:

$$\nabla I = (I \otimes h_x)\hat{x} + (I \otimes h_y)\hat{y}, \qquad (3.14)$$

where $\otimes$ is the convolution operator.

Several gradient filters exist and some examples are given in [98]:

- Basic filters (respectively forward and central differences):

$$h_x = h_y^T = \begin{bmatrix} 1 \\ -1 \end{bmatrix}, \quad h_x = h_y^T = \begin{bmatrix} 1 \\ 0 \\ -1 \end{bmatrix}. \tag{3.15}$$

- Sobel filter:

$$h_x = \begin{bmatrix} 1 & 0 & -1 \\ 2 & 0 & -2 \\ 1 & 0 & -1 \end{bmatrix}, \quad h_y = \begin{bmatrix} 1 & 2 & 1 \\ 0 & 0 & 0 \\ -1 & -2 & -1 \end{bmatrix}. \tag{3.16}$$

Note that these equations work well for gray-scale images, but they are rough approximations for color images [69]. Two important measures can be computed from the gradient—its magnitude and directions, which are respectively defined as

$$\|\nabla I\| = \sqrt{(I \otimes h_x)^2 + (I \otimes h_y)^2}, \tag{3.17}$$

$$\|\nabla I\|_\theta = \arctan\left(\frac{I \otimes h_y}{I \otimes h_x}\right). \tag{3.18}$$

A color image can be directly converted into a gray-scale image by ma-

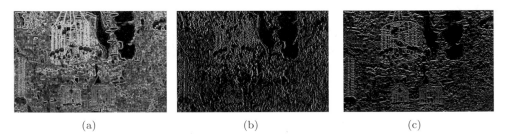

| (a) | (b) | (c) |

Figure 3.4. Gradient magnitude results using a Sobel filter on the channel R of an RGB image (a) and its directional gradient ∇R_x (b) and ∇R_y (c).

nipulating its gradient. Gooch et al. [99] tested a strategy to overcome the isoluminant colors issue, augmenting the luminance gradient with the gradient of the chrominance channels. Starting from the CIE $L^*a^*b^*$ version, of the input color image, the gradient for L^*, a^*, and b^* are computed in the two directions x and y as ∇L_i, ∇a_i, and ∇b_i, respectively. The i index indicates a direction, either x or y.

For both directions, x and y, a Gaussian weighting is used to modify the luminance gradient as follows:

$$\nabla Y_i' = \left(\exp(-w \cdot \nabla L_i^2) \cdot q\right) + \nabla Y_i, \tag{3.19}$$

where ∇Y_i is the gradient of the starting gray-scale image taken as the luminance value Y, and the q factor is computed as

$$q = sign(\nabla Y_i)\sqrt{\nabla a_i^2 + \nabla b_i^2}. \qquad (3.20)$$

A Poisson solver is then applied on the new gradient for reconstructing the final gray-scale image [225]. Other types of gradient manipulation can be employed, to extract the gray-scale image from the input color image. An example is shown in Figure 3.5, where gray-scale images are obtained with the maximum gradient operator Figure 3.5 (a) and Gooch et al.'s gradient method [99] Figure 3.5 (b).

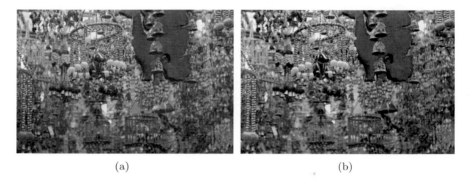

(a)   (b)

Figure 3.5. A gray-scale image is obtained with two types of gradient image manipulation. The input color image is the one used in Figure 3.3. A gray-scale image obtained using the maximum gradient operator (a). Gray-scale image obtained using Gooch et al.'s gradient method [99] (b).

More sophisticated approaches that use the image gradient to convert a color image into a gray-scale image have been proposed [141,199]. Neumann et al. [199] proposed a gradient field computation formula that takes advantage of Coloroid color space. The gradient is measured as color difference, which includes the contribution of luminance Y, saturation s, and hue h. Kim et al. [141] proposed a simple nonlinear mapping function of the form

$$g(x,y) = L + f(\theta)C. \qquad (3.21)$$

As a first step, the input color image is converted to the CIE $L^*C^*h^*$ color space, and then Equation 3.21 is applied. The hue h is expressed in its angle form θ. The function f is modeled as [141]

$$f(\theta) = \sum_{k=1}^{n}(A_k \cos(k\theta) + B_k \sin(k\theta)) + A_0. \qquad (3.22)$$

The unknown parameters A_k, B_k, and A_0 are obtained through the minimization of an objective function, through a nonlinear global mapping strategy [141]. This allows maintaining features discriminability, as the local differences between pixels, into the gray-scale image. The image gradient is used for representing local pixel differences, where the normalized color distance between two pixels is computed in the CIE $L^*a^*b^*$ color space. The objective function, E, is expressed as the difference of the image gradient between the gray-scale output image g and the input color image as

$$E = \sum_{(x,y) \in \Omega} ||\nabla g - D(x,y)||^2. \tag{3.23}$$

The gradient of the gray-scale image is computed using forward differences (the first filter in Equation 3.16). D is defined as the difference of input pixel color c as

$$D(x,y) = \begin{bmatrix} c(x+1,y) - c(x-1,y) \\ c(x,y+1) - c(x,y-1) \end{bmatrix}^T, \tag{3.24}$$

where the difference operator is defined as follow

$$c_i - c_j = sign(c_i, c_j) \sqrt{\Delta L_{ij}^2 + \left(\alpha \frac{\sqrt{\Delta a_{ij}^{*2} + \Delta b_{ij}^{*2}}}{R} \right)^2}, \tag{3.25}$$

where R is a normalization constant used to equalize the dynamic range of the chromatic and lightness contrast; it is equal to $2.54\sqrt{2}$. α is a user-defined parameter that controls the influence of chromatic contrast on features discriminability [141]. The ΔK_{ij} identifies the difference between the color space components of pixels i and j. The *sign* function encodes the relative ordering of two pixels. Kim et al. [141] used a lightness predictor approach similar to Smith et al. [251].

3.5   Minimization Problem

As shown in Section 3.2.1, a simple conversion of the RGB color input space into a luminance value Y, e.g., Equation 3.3, leads to the loss of several details. To see if this drawback can be solved, we can use either conversion equations which take into account color appearance correlates or a different color space conversion. Figure 3.6 shows results of alternative equations, using either lightness L^* or luminance Y. However, these solutions none of preserve the full details and relevant information that is visible in the input RGB color image. To solve the above drawback, Gooch et al. [100] proposed

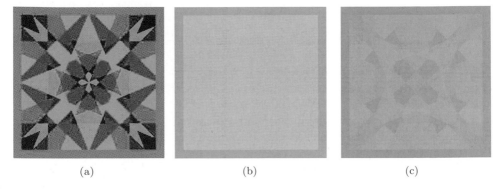

(a)   (b)   (c)

Figure 3.6. Color2gray mapping using either lightness L^* or luminance Y channel. (a) input color RGB image, (b) gray-scale image using L^* from the CIE $L^*a^*b^*$ and (c) gray-scale image using Y from the YC_bC_r color spaces. Input image courtesy and copyright of Laszlo Neumann.

a method where the differences in the input color image are encoded into luminance differences in the gray-scale image. The target difference, δ_{ij}, is defined as the difference between the pixel i and its neighbor pixel j. Since the image pixels are represented in the CIE $L^*a^*b^*$ color space, we have two types of differences; the $\Delta L^*_{ij} = L^*_i - L^*_j$ for the lightness, and the $\Delta \vec{C}_{ij}$ for chrominance. The chrominance difference is a vector that is based on the differences of the a^* and b^* chrominance channels $(\Delta a^*_{ij}, \Delta b^*_{ij})$. Since two types of differences exist per pixel, the predominant one needs to be selected. First, the chrominance difference, which is a $2D$ vector, is mapped to a single dimension using the Euclidean norm $(\|\Delta \vec{C}_{ij}\|)$ and a parameter θ is introduced to parametrize its sign. The target difference δ_{ij} is defined as follows

$$\delta(\alpha, \theta)_{i,j} = \begin{cases} \Delta L^*_{ij} & \text{if } |\Delta L^*_{ij}| > crunch(\|\Delta \vec{C}_{ij}\|) \\ crunch(\|\Delta \vec{C}_{ij}\|) & \text{if } \Delta \vec{C}_{ij} \cdot \vec{v}_\theta \geq 0 \\ crunch(-\|\Delta \vec{C}_{ij}\|) & \text{otherwise} \end{cases}, \quad (3.26)$$

where the *crunch* function and $\vec{v}_\theta$ are respectively defined as

$$crunch(x) = \alpha \tanh(x/\alpha), \qquad (3.27)$$

$$\vec{v}_\theta = (\cos \theta, \sin \theta), \qquad (3.28)$$

where θ and α are user-controlled parameters: θ determines whether chromatic differences are mapped to increases or decreases in luminance value, and α controls the amount of variation that is allowed to change the source luminance value [100]. The final gray-scale image is obtained minimizing the following objective function:

$$f(g) = \sum_{(i,j)\in K} \left((g_i - g_j) - \delta_{ij}\right)^2, \qquad (3.29)$$

where K is a set of ordered pixel pairs (i, j). The idea is to initialize g with the luminance channel of the input color image, and then use a conjugate gradient descended method [225] to find the minimum of the function Equation 3.29. The Rasche et al. [228] color2gray mapping method is based on the same principle as Gooch et al. [100], where the perceived color differences between any pair of colors need to be proportional to their perceived gray difference. The two approaches are quite similar, and the differences lie in the color difference computation. In particular, Rasche et al. [228] do not separate the lightness difference from the chrominance difference, so avoiding strategies to define which one is predominant. Moreover, the color difference is computed using the well-known $CIE94$ color difference metric [61]. Similar to Gooch et al. [100], a conjugate gradient descended method, Fletcher–Reeves [225], is used to minimize the error objective function.

3.6 Predominant Component Analysis

In Section 3.3, we have seen that the main characteristic of the PCA is to optimize the variability of observations, finding the component of the 3-dimensional input color space that is carrying the most important information of the entire color space. On the other hand, predominant component analysis optimizes the differences between observations [103]. Grundland et al. [103] share the main objective of several color2gray mapping methods, which is to compensate for color contrast information lost in the luminance channel. As explained in the above sections, this is achieved by finding a single chromatic coordinate that is carrying this information and add it back to the luminance channel. To achieve this goal, Grundland et al. [103] make use of predominant component analysis to develop a framework that can be separated into five steps and take in as input a linear RGB color image:

- YPQ **color space**: YPQ color space has been chosen for its computational simplicity while keeping its intuitive interoperation of its components, where Y is an achromatic luminance channel, while P

(yellow-blue) and Q (red-green) are the chromatic channels:

$$\begin{bmatrix} Y \\ P \\ Q \end{bmatrix} = \begin{bmatrix} 0.2989 & 05870 & 0.1140 \\ 0.500 & 0.500 & -1.000 \\ 1.000 & -1.000 & 0.000 \end{bmatrix} \begin{bmatrix} R \\ G \\ B \end{bmatrix}. \tag{3.30}$$

It worth to notice that the Y channel is computed using the standard ITU Rec.601 [129] conversion equation from RGB to luminance Y. The saturation s and the hue h can be easily derived as follows

$$s = \sqrt{P^2 + Q^2}, \tag{3.31}$$

$$h = \frac{1}{\pi} \tanh \left(\frac{Q}{P} \right). \tag{3.32}$$

- **Image sampling**: This is needed to analyze the distribution of color contrast between image features. Traditional techniques based on neighborhood pixels or surrounding regions, blur color contrast and they are computationally expensive. To overcome these issues, Grundland et al. [103] introduced a Gaussian pairing strategy, where each image pixel x_i is paired with a pixel x_i'. The x_i' is chosen randomly using an isotropic bivariate Gaussian distribution [103].

- **Predominant component analysis**: The paired pixels are used to compute the color differences D_i, in the RGB color space. Then, this is used by the predominant component analysis, to find the chromatic channel that carries the contrast lost in the luminance channel. The color differences are computed with the classical Euclidean distance. From these, the contrast loss c_i, incurred when the luminance difference ΔY_i is used to represent the RGB color difference D_i, is computed as

$$c_i = \frac{D_i - (1/Y_{length})|\Delta Y_i|}{D_i}, \tag{3.33}$$

where Y_{length} measures the length of the Y axis, which is equal to 0.6686. Afterward, on the PQ plane, the predominant axes of chromatic contrast, Δp and Δq, are computed as

$$\Delta p = \sum_i o_i c_i \Delta P_i, \tag{3.34}$$

where the ΔP_i is the difference between the paired pixels x and x_i for the channel P, and o_i is the sign of the luminance difference that defines the luminance ordering. This allows us to keep the same ordering in the color contrast channel and avoid artifacts due to contrast reversal. For the component Q, a similar equation to Equation 3.34 is used, where the channel P_i is substituted with the channel Q_i.

- **Combining luminance and chrominance**: A linear combination
  is used to merge luminance and color contrast information. As the
  first step, the chrominance values are projected on the predominant
  axes Δp and Δq, obtaining the predominant chromatic channel C_i.
  Then, the luminance values Y_i and the predominant chromatic channel
  C_i are merged as

$$Y_i' = Y_i + \lambda C_i, \tag{3.35}$$

  where λ is an enhancement factor with a typical value of 0.3.

- **Dynamic range adjustment**: The final step adjusts the Y_i' dynamic
  range and compensates for the image noise [103].

3.7   Lightness Predictor

As seen in Chapter 2, color appearance models are able to predict the
human perceptual response to color stimulus. An interesting appearance
phenomenon is the so-called Helmholtz–Kohlrausch (HK) effect. Taking a
chromatic stimulus with the same luminance as a white reference stimulus, it
will appear brighter than the reference [251]. This phenomenon makes clear
that not only is luminance Y contributing to the perception of lightness,
but also the chroma component C of a color stimulus is contributing to it.
This varies in function of luminance and hue h [251]. In other words, the
HK effect is predicted introducing a chroma light term that corrects the
lightness L^* [251]. This term is based on the chromatic color's components.
Different approaches exist to compute the HK effect:

- **Fairchild et al. CIE $L^*a^*b^*$ based metric** [77]

$$L^{**} = L^* + (2.5 - 0.025L^*) \left(0.116 \left| \sin \left(\frac{h - 90}{2} \right) \right| + 0.085 \right) + C. \tag{3.36}$$

- **Nayatani Variable Achromatic Color** [197, 198]

$$L^{**} = L^* + [-0.1340q(\theta) + 0.0872K_{Br}]s_{uv}L^*. \tag{3.37}$$

- **Nayatani Variable Chromatic Color** [197, 198]

$$L^{**} = L^* + [-0.8660q(\theta) + 0.0872K_{Br}]s_{uv}L^*. \tag{3.38}$$

For the Fairchild et al. model [77], L^* is the lightness, h is the hue,
and C is the chroma, while for both Nayatani models the parameters are
derived from the CIE $L^*u^*v^*$ color space. The u' and v' components are

the chromaticities of the color stimulus, u'_n and v'_n are the chromaticities of
the reference white, and L_a is the adapting luminance set to 20 cd/m^2 in
Nayatani works [197, 198]:

$$\theta = \arctan\left(\frac{v' - v'_n}{u' - u'_n}\right), \tag{3.39}$$

$$\begin{aligned} q(\theta) = &-0.01585 - 0.03017\cos(\theta) - 0.04556\cos(2\theta) \\ &-0.02667\cos(3\theta) - 0.00295\cos(4\theta) \\ &+0.14592\sin(\theta) + 0.05084\sin(2\theta) \\ &-0.01900\sin(3\theta) - 0.00764\sin(4\theta), \end{aligned} \tag{3.40}$$

$$K_{Br} = 0.2717\frac{6.469 + 6.362L_a^{0.4495}}{6.469 + L_a^{0.4495}}, \tag{3.41}$$

and the saturation parameter s_{uv} is defined as

$$s_{uv} = 13\sqrt{(u' - u'_n)^2 + (v' - v'_n)^2}. \tag{3.42}$$

The color2gray approach based on a lightness predictor [251] uses the above
lightness predictor model (Equation 3.37) to partially solve the problem of
color ordering typical of luminance mapping. However, chromatic contrast
may be overenhanced and it has to be reduced. To solve this issue, Smith
et al. [251] proposed a local contrast adjustment making use of Laplacian
pyramids. As the first step, both the input color image I and its gray-
scale g (based on the lightness predictor; Equation 3.37), are converted
to CIE $L^*a^*b^*$. Then, a four-level $\mathbf{h}^i$ Laplacian pyramid is built for each
image. The gray-scale image g_{L^*} is modified as follows

$$g'_{L*} = g_{L^*} + \sum_{i=1}^{n-1} k_i\lambda_i\mathbf{h}_i\{g_{L^*}\}, \tag{3.43}$$

where n is the number of levels in the pyramid, $k_i \leq 1$ values control the
spatial effect [251], and λ_i values are gains measuring the amount of contrast
needed to match the color contrast $h_i(I)$ of the input image at level i. Note
that the value λ_i is increased at each level i, and it is defined as

$$\lambda_i = \left(\frac{\Delta E(\mathbf{h}_i\{I\})}{|\mathbf{h}_i\{g_{L^*}\}|}\right)^p, \tag{3.44}$$

where p remaps λ to a nonlinear scale to avoid overemphasizing stronger
contrast, and ΔE is a measure of the color contrast calculated as the color
difference between a pixel and its neighborhood.

3.8   Image Fusion

Image fusion (see Section 2.7.1) has been used extensively in image processing and computer graphics techniques such as HDR imaging [185], image editing [216], etc. It is based on the idea that several images with different features can be combined in one image [10], retaining, based on the specific application, the most important features. Ancuti et al. [10] proposed a framework based on the fusion concept; see Figure 3.7.

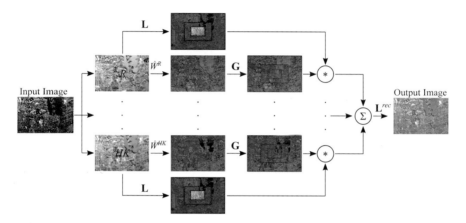

Figure 3.7. Color2gray framework proposed by Ancuti et al. [10]. First, the input color image is separated in four components: R, G, B, and its HK lightness predictor based on the Fairchild et al. model [77]. Second, a Laplacian pyramid is built for each component. Third, a weight map and its Gaussian pyramid are computed for each component. Finally, the fuse pyramid is computed, and an output gray-scale image is reconstructed from it.

The framework has four main steps:

- **Input Separation**: The input color image is separated in four components ($K = 4$): R, G, B, and its HK based lightness predictor in Equation 3.36 [77].

- **Weights Computation**: Three weight maps are computed for each component. A saliency weight map, W_s, tries to preserve the salient regions where the HVS is mainly focused. At pixel (x, y), W_s is defined as

$$W_s(x, y) = \|I_g^k - I_b^k(x, y)\|, \qquad (3.45)$$

where k is the index of a component, and I_g^k and I_b^k are, respectively, the arithmetic mean and the blurred version of I^k.

To overcome perception degradation in over/under exposed regions, an exposedness weight map, W_e, is similarly computed. This is defined as in Mertens et al.'s work [185] as

$$W_e(x, y) = \exp\left(-\frac{(I^k(x, y) - 0.5)^2}{2 \cdot (0.25)^2}\right), \qquad (3.46)$$

where $I_k(x, y)$ is the pixel of the input image I^k. Note that this is an estimation of how a pixel is well exposed, and it tries to maintain the appearance of the local contrast [10]. Finally, a chromatic weight map, W_c, is computed for balancing the influence of chromatic stimuli into the perception of lightness [10]. W_c is a function of the saturation s appearance correlate. In particular, it is a standard deviation measure between each input and the saturation s of the input color image [10]. These three maps are then multiplied together obtaining a final weight map W for each component; $W = W_s \times W_e \times W_c$.

- **Multi-scale Decomposition**: Each component is decomposed into a Laplacian pyramid, and each normalized weight map into a Gaussian pyramid. Both pyramids have the same number of levels i. The use of a multi-scale decomposition captures salient information at different scales.

- **Image Fusion**: The fused pyramid version $\mathbf{L}^i\{F\}$ is obtained as

$$\mathbf{L}^i\{F\} = \sum_{k=1}^{K} \mathbf{G}^i\{\hat{W}^k\} \times \mathbf{L}^i\{I^k\}, \qquad (3.47)$$

where $K = 4$ is the number of input components, $\mathbf{G}^i\{\hat{W}^k\}$ is the i level of the Gaussian pyramid of the normalized weight map, and $\mathbf{L}^i\{I^k\}$ is the Laplacian pyramid level of the input component I^k.

3.9   Filtering

Typically, lost information during the color2gray mapping is evident around edges between image areas with similar luminance; i.e., a typical case of isoluminant colors. Edges are either partially or totally lost depending on how similar the luminance is between adjacent image areas. High-pass filters have been used in image processing and computer graphics applications to preserve edges. Their use can also be exploited in the color2gray mapping, where high frequency chrominance information is added to the luminance channel. Bala et al. [22] first proposed this idea, which shares two main aspects that are common in many of the color2gray mapping methods presented in this chapter so far. The first, is that luminance and chrominance information may have different polarity, which means that combining them will not contribute to enhancing the luminance information. The second is that the chrominance information needs to be added only in image areas where the luminance information is not sufficient to discriminate the differences between these two areas. How can we properly address these two important issues? Figure 3.8 shows the framework proposed by Bala et

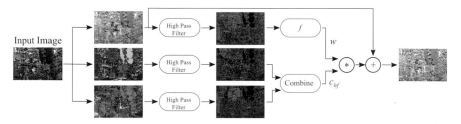

Figure 3.8. Framework of the color2gray mapping proposed by Bala et al. [22].

al. [22]. Having as input a CIE $L^*a^*b^*$image, a high-pass filter is applied to the L^*, a^*, and b^* channels. In this way, high-frequency information can be extracted and further manipulated. Afterward, a_{hf}^* and b_{hf}^* high-frequencies are combined in one chrominance signal either with the classical Euclidean or 1-norm metrics (c_{hf}). A weighting function is computed to combine the lightness and chrominance contributions:

$$w = sign(L_{hf}^*)f(|L_{hf}^*|), \qquad (3.48)$$

where L_{hf}^* is the high-frequency lightness information. The *sign* takes into account the differences in polarity between luminance and chrominance. The function f addresses the need to add chrominance only in areas where the predominant information is not only carried by the luminance, but also by chrominance. In the work of Bala et al. [22], the choice of function f relies on the application where the algorithm will be used; i.e., printing. However, any other type of function can be used to convey the wanted effect. The high-pass filter is defined as the difference between the color image and an $N \times N$ box filter. For example, L_{hf} for the luminance channel is computed as

$$L_{hf}^*(i) = L^*(i) - \frac{1}{N^2} \sum_{j \in S(i)} L^*(j), \qquad (3.49)$$

where $S(i)$ is an $N \times N$ neighborhood around the pixel i [22]. Similar equations are used for the a^* and b^* channels for computing a_{hf}^* and b_{hf}^*. Finally, the lightness value of the input color image L^* is modified as follows to take into account the chrominance contribution:

$$L_{out}^*(i) = L^*(i) + c_{hf} * w. \tag{3.50}$$

This operation shares some similarities with unsharp masking, but the high-pass filter component comes from the chromatic channels instead of a usual high-pass filter enhancement directly in the luminance channel. This can be seen as an extended version of the Cornsweet illusion. In fact, Smith et al. [251] employ a similar strategy but their approach is multi-resolution where the spatial range of the high-pass filter is adaptively driven by λ_i as shown in Equation 3.44.

3.10   Colorization: The Opposite Problem

Colorization can be seen as part of the large category of problems named color mapping [83]; that also includes the so-called color transfer problem. In the latter case, a color input image is re-colored based on some goals. In this section, we introduce the colorization problem, while the color transfer problem will be treated in Section 4.2.

Colorization involves starting from a gray-scale input image (1-dimensional space), and reconstructing a 3-dimensional space (color image), based on specific constraints. Since there is not a unique solution to this problem, this is an ill-posed problem. Typically, the input is a gray-scale image with luminance information $Y(x, y)$, and a reference color image (see Figure 3.9) or a color user-based stroke images $U(x, y)$ and $V(x, y)$ (chrominance) (see Figure 3.10) [85].

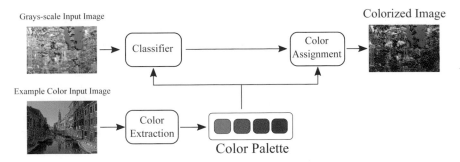

Figure 3.9. Starting from a gray-scale and reference color images as input, color categories can be extracted and further used to color the gray-scale input image.

A naïve approach to solving the colorization problem is to manually select 255 color patches and use them to color the gray-scale input image [98]. In this approach, the goal is not to produce realistic looking color images [2], which may be seen as strong constraint that may limit its uses

Input Image   Colorized Image

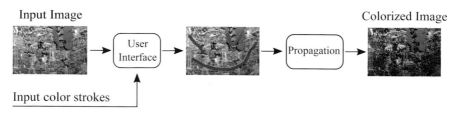

Input color strokes

Figure 3.10. Starting from a gray-scale image and color strokes information as input, the end-user can apply the color strokes on some areas of the gray-scale input image and then this information is propagated.

in several applications. If the goal is to produce realistic color images, more sophisticated techniques need to be adopted. First, these techniques need to be able to select the best color candidates from a large set of colors provided as reference. Second, this process should be automated as much as possible.

Welsh et al. [288] proposed a colorization technique that jitters a number of colors from the reference color image and uses this subset for the colorization process. The mapping is done by looking at each pixel of the gray-scale input image and finding the best match with the subsample colors. The match is found using a weighted average of the gray-scale pixel and its neighborhood statistics. The neighborhood consists of a 5×5 region, and the standard deviation is used as the statistic. The chromatic values of the color sample $\alpha\beta$, in color space $l\alpha\beta$, are added to the original gray-scale pixel, retaining its original luminance value. However, if regions in the gray-scale image have largely different luminance values, then in the source image, the results may show artifacts [288]. To solve this problem, and to increase user interaction, Welsh et al. [288] introduced the use of swatches between corresponding regions (reference colors and gray-scale images). As will be introduced in Chapter 4, the swatches are small areas selected by the end-user that delimit the part of the image where the mapping operation will be performed. This allows selecting areas, with very limited intensity discontinuities between the reference color image and the input gray-scale image. Levin et al. [156] introduced a constraint where two neighboring pixels, a and b, should have similar chrominance if they have similar intensities. This can be obtained by minimizing the chrominance difference between the chrominance of pixel a and the weighted sum of the chrominance of the neighboring pixel b (for chrominance U):

$$E(U) = \sum_{x_a, y_a} \left(U(x_a, y_a) - \sum_{(x_b, y_b) \in N(x_a, y_a)} wU(x_b, y_b) \right)^2, \qquad (3.51)$$

The same Equation 3.51 is also used for the chrominance V. w is a weighting function that sums to one. A possible choice is the sigmoid function of the squared difference between the two intensities of the two neighboring pixels. Also a weighting function based on the normalized correlation between the two intensities of the two neighboring pixels has been tested [156]. Given a set of location selected by the end-user through the use of strokes, a_i, the objective function of Equation 3.51 is minimized for both chrominance U and V. This is a large linear system that can be solved with a traditional minimization technique for linear problems. However, the linear system can be quite large and is often poorly preconditioned, making the problem difficult to solve and time consuming. Wavelet-based edge-aware interpolation techniques [13, 85], applied to the colorization problem may solve the above issue. This type of wavelet is based on a lifting scheme that provides a simple and efficient framework to construct a multi-scale edge-aware wavelet [85]. A PCA-based colorization method has been presented by Abadpour et al. [2]. This method allows the end-user to define an index map to select the class where each pixels belongs, and to select a homogeneous color swatch as the reference color from which the color information is extracted and applied to the respective swatches of the gray-scale image. The process is repeated for all the swatches defined by the end-user.

One of the major drawbacks of the above methods is that considerable effort is required by the end-user. On one hand, this helps to achieve qualitatively better results. On the other hand, it makes the method tedious and not fully automatic. A second drawback is that images may look unnatural because of limited user skills or inaccurate color mapping [52]. Chia et al. [52] proposed a method that reduces the end-user intervention to segment the input gray-scale image into background and foreground objects by providing a semantic text label for each object. These labels are used to download a large set of photos from image-sharing websites such as Flickr, Google Image Search, and Picasa. Ten downloaded images are used to find the most suitable reference photo for both background and foreground objects, by filtering the search results with respect to their similarity to the gray-scale input image. The colorization is performed by minimizing an energy function. Through a user interface, various solutions are shown to the end-user, who selects the most appropriate one.

3.11 Summary

Color2gray is a challenging and difficult problem, where all relevant information contained in three color channels is mapped into a single channel. A list of goals and aims have been listed in this chapter, followed by a classification of several methods which try to address, either partially or in full, these goals. Despite this long list of methods, other methods also exist that propose improvements in some of the steps of a general color2gray approach [58, 73]. In particular, Drew et al. [73] proposed a method to solve the sign luminance problem, while Cui et al. [58] introduced the use of ISOMAP [265] to find the pairwise distance between all the pairs of color pixels. Despite the availability of this huge number of methods, a complete evaluation does not exist. However, Čadik [43] has been the first to provide a meaningful evaluation of a large number of color2gray methods. He ran two subjective experiments in which a total of 24 color images were converted to gray-scale using seven color2gray methods and evaluated by 119 subjects using a paired comparison approach. His main goal was to understand the strengths and weaknesses of selected methods to be used as input for the development of a new color2gray conversion.

4

Style Retargeting

4.1 Introduction

Recently, transferring attributes such as colors, tones, contrast, photographic look, etc., have become extremely important. First, the color adjustment in a photograph can take a lot of time, and repeating it for a whole album of photographs in a coherent way can be very tedious and may take hours. Second, app users, who have been empowered by modern mobile devices with cameras want to have special looks, be original, and be unique, by copying other people's styles. Third, automation of the grading pipeline can save a lot of time for artists, avoiding tedious tasks when defining the style and the mood for thousands of images.

4.2 Color Transfer

Pure color transfer together with color2gray and colorization (see Chapter 3) is a color mapping problem [83] whose goal is to find a mapping, T, for transferring colors from a target image, I_t, onto a source image, I_s; see Table 4.1. Typically, this problem is formally defined as

$$I_o = T(I_s, I_t). \tag{4.1}$$

Note that both I_s and I_t are color images. To find a high-quality T is challenging. This is because the mapping has to deal with 3D color distributions, and it has to be tolerant of compression artifacts, noise, dark images (where colors are not too distinguishable), and to be spatially coherent.

4.2.1 Histogram Matching

A classic image processing technique for color transfer from a target image to a source image is histogram matching [98]. Assuming gray-scale images,

Figure 4.1. The general concept of color and style retargeting: Colors and style of a given target image are transferred to the source image while keeping its structure and composition.

Symbol	Meaning
s	for referring to a source image or property
t	for referring to a target image or property
o	for referring to a re-targeted source image or property
I_s	a source image
I_t	a target image
I_o	a re-targeted source image

Table 4.1. This table shows the main symbols used in this chapter.

a histogram H of an image, I, is typically formally defined as

$$H = \Big(|A_1|, ..., |A_n| \Big),\tag{4.2}$$

where an A_j is defined as

$$A_j = \Big\{ k | I(k) \in \big[\Pi(j-1), \Pi(j) \big] \Big\},\tag{4.3}$$

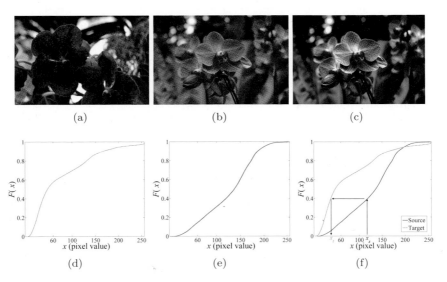

Figure 4.2. An example of histogram matching: (a) The target image. (b) The source image. (c) The result of histogram matching. (d) The CDF of the green color channel for (a). (e) The CDF of the green color channel for (b). (f) Histogram in (d) and (e) in the same plot showing how histogram matching works.

and Π, a normalization operator, is defined as:

$$\Pi(x) = x \frac{I_{\max} - I_{\min}}{n} + I_{\min}. \tag{4.4}$$

In order to match the appearance between two images, I_t and I_s, their histograms, H_t and H_s, need to be computed; see Equation 4.2. From these, the cumulative distribution functions (CDFs), F_s and F_t, are calculated; a CDF is typically defined as

$$F(x) = \frac{\sum_{i=1}^{x} H(i)}{\sum_{i=1}^{n} H(i)}. \tag{4.5}$$

Examples of CDFs are shown in Figure 4.2 (a) and Figure 4.2 (b). Then, each value of the source image, x_s, is remapped to an x_t value such that $F_s(\Pi_s(x_s))$ is equal to $F_t(\Pi_t(x_t))$; see Figure 4.2 (c). This can be formally written as

$$x_t = T(x_s) = F_t^{-1}(F_s(\Pi_s(x_s))). \tag{4.6}$$

In the case of color images, this process is repeated for each color channel. Note that results can vary depending on the used color space. For example, Figure 4.3 shows how the result varies with different color spaces. Note that the use of the RGB color space typically gives unpredictable results because its axes are highly correlated and perceptually nonuniform. Histogram matching can introduce color distortions, which can be alleviated by histogram warping [103]. This method employs classic histogram matching for establishing a mapping, Equation 4.6, between the corresponding set of quantiles of I_s and I_t. These are used as parameters for piecewise rational quadratic interpolating splines; i.e., the warping transformation T for remapping values.

(a) RGB (b) CIE $L^*a^*b^*$

Figure 4.3. An example showing histogram matching in different color spaces using the input images in Figure 4.2.

One issue of histogram matching is that source gradients can be greatly modified without any preservation. To overcome this issue, gradients can be enforced after color transfer [299]. So, histogram matching is first applied to a source image, I_s, using a target image, I_t, obtaining an intermediate image, I_f. Second, the final image, I_o, is computed by minimizing differences between the original gradients and transferred colors. This minimization is defined as

$$\arg\min_{I_o} \sum_i \left(I_o(i) - I_f(i) \right)^2 + \lambda \sum_j \left(\frac{\partial I_o(j)}{\partial x} - \frac{\partial I_s(j)}{\partial x} \right)^2 + \left(\frac{\partial I_o(j)}{\partial y} - \frac{\partial I_s(j)}{\partial y} \right)^2,$$

$$(4.7)$$

where the first term is the *color term*, which is responsible for transferring colors, and the second term is the *gradient term*, which keeps gradients of the input image. λ is a coefficient for weighting the importance of gradient preservation and target colors; experimental results showed that $\lambda = 1$ works well for most images. An example of this method is shown in Figure 4.4.

Another important issue of classic histogram matching is the lack of artistic control and support for HDR images. For example, an artist should be allowed to partially transfer colors from a target to a source image.

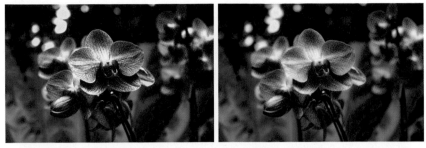

(a) without gradient enforcement   (b) with gradient enforcement

Figure 4.4. An example of gradient enforcement after color transfer between images [299] using the input images in Figure 4.2. Note that spatial coherency and fine details in (b) are better preserved than in (a).

Pouli and Reinhard [223, 224] introduced a progressive transfer scheme that provides a solution to these issues. This method reshapes the source histogram, H_s, to match the target one, H_t, obtaining H_o. The reshaping works at different scales, up to $S_{\max}$ scales, and the user can decide the matching percentage. For example, a 30% matching means that H_o will be processed using only the first $\lceil 0.3 S_{\max} \rceil$ scales. The reshaping at each scale is achieved by transferring the mean, μ, and standard deviation, σ, between matched regions (detected by minima) of source and target histograms at a given scale, k:

$$\forall i \in R_{j,k} \quad H_{o,k}(i) = \left(H_{s,k}(i) - w_k \mu_{s,k}(j) \right) \frac{(1 - w_k)\sigma_{t,k}(j)}{w_k \sigma_{s,k}(j)} + (1 - w_k)\mu_{t,k}(j),$$

$$(4.8)$$

where $R_{j,k}$ is a matched region of the histogram at scale k, and $w_k = k/S_{\max}$ controls the amount of matching between the I_s and I_t. Then, classic histogram matching is applied to I_s using H_o for each color channel in the CIE $L^*a^*b^*$ color space (this provides a basic TMO for HDR images; i.e., the cube root) obtaining I_o. This method also provides a mechanism for gradient preservation based on transferring details extracted using the bilateral filter; see Appendix A. This mechanism is defined by the following equations:

$$I'_{\text{detail},o} = I_o + w_c \left(D(I_s) - D(I_o) \right) \quad \text{and} \quad D(I) = I - BF[I], \qquad (4.9)$$

where $BF[\cdot]$ is a bilateral filter and w_c is a parameter that controls the amount of correction to apply. Experimental results showed that $w_c = 1$ works when noise is present in smooth areas of LDR images, and $w_c = cS_{\max}$ for HDR images where c is a contrast parameter. Figure 4.5 shows an example of this technique.

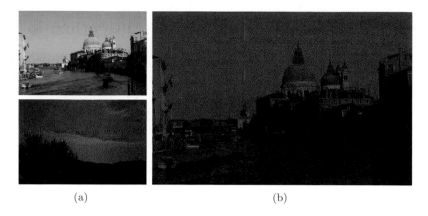

(a)   (b)

Figure 4.5. An example of progressive transfer [223, 224]: (a) At the top, I_s; and at the bottom I_t. (b) The result of the color transfer.

3D Histogram Matching. Matching 3D color histograms can improve quality results in color transfer when compared to three separate one-dimensional histogram matchings. However, matching 3D distributions is a computationally complex problem to solve.

To reduce the problem complexity, an idea is to project these histograms onto a 1D axis to determine a mapping, f, between these projections, and to apply f to the original 3D samples of I_s [218, 219]. In more in detail, the process is defined by the following steps:

1. A random rotation $\mathbf{R}$ is generated.

2. Colors of I_s and I_t are rotated using $\mathbf{R}$; $\hat{I}_s = \mathbf{R}I_s$ and $\hat{I}_r = \mathbf{R}I_r$, and then projected onto each axis.

3. For each axis, a transformation t_i is computed and applied to each color component; given $\hat{I}_s = (c_1, c_2, c_3)$ the application is defined as $\hat{I}_s' = (t_1(c_1), t_2(c_2), t_3(c_3))$.

4. $\hat{I}_s'$ colors are rotated back; $I_s' = \mathbf{R}^{-1}\hat{I}_s'$.

This process is repeated until convergence of all marginals is reached for every possible rotation. In practice, this process typically converges after 20 iterations. An example of this technique is shown in Figure 4.6.

Another solution to approximate the 3D matching problem is to apply a sequential sampling technique in the HSL color space [200].

(a)   (b)

Figure 4.6. An example of the 3D histogram matching applied to color transfer method [219]: (a) At the top, I_s; and at the bottom, I_t. (b) The result of the color transfer.

4.2.2 Statistical Transfer

Color transfer can be also achieved by applying linear transformation to color values. The seminal work by Reinhard et al. [229] introduced this methodology of superimposing statistics of the target color channels such as mean and standard deviation. Assuming input images, a target and a source in the RGB color are first converted into the XYZ color space; a modified RGB2XYZ ITU Rec. 601 color matrix is used to force the white point chromaticities to have coordinates $x = 0.33$ and $y = 0.33$. Then, colors are converted into the LMS color spaces; see Chapter 2. Given that XYZ and LMS are linear color spaces, this operation can be done with a single color matrix conversion (i.e., the concatenation of two matrices) which is defined as

$$\begin{bmatrix} L \\ M \\ S \end{bmatrix} = \begin{bmatrix} 0.3811 & 0.5383 & 0.0402 \\ 0.19667 & 0.7244 & 0.0782 \\ 0.0241 & 0.1288 & 0.8444 \end{bmatrix} \begin{bmatrix} R \\ G \\ B \end{bmatrix}. \tag{4.10}$$

Second, LMS images are converted into the $l\alpha\beta$ color space which has the property to decorrelate color axes for natural images [239]. This is achieved by applying the $l\alpha\beta$ conversion matrix to LMS values:

$$\begin{bmatrix} l \\ \alpha \\ \beta \end{bmatrix} = \begin{bmatrix} \frac{1}{\sqrt{3}} & 0.0 & 0.0 \\ 0.0 & \frac{1}{\sqrt{6}} & 0.0 \\ 0.0 & 0.0 & \frac{1}{\sqrt{2}} \end{bmatrix} \begin{bmatrix} 1.0 & 1.0 & 1.0 \\ 1.0 & 1.0 & -2.0 \\ 1.0 & -1.0 & 0.0 \end{bmatrix} \begin{bmatrix} \log L \\ \log M \\ \log S \end{bmatrix}. \tag{4.11}$$

Note that LMS values may have a large skew which can be mitigated by applying the logarithm to them before $l\alpha\beta$ color space conversion. Third,

(a)   (b)

Figure 4.7. An example of Reinhard et al.'s method [229]: (a) At the top, I_s; and at the bottom, I_t. (b) The result of the color transfer.

statistics are transferred in the $l\alpha\beta$ color space. For each image and each color channel, mean values, μ, and standard deviation values, σ, are computed. At this point, source statistics are substituted with target statistics using a *shifting, scaling, and then shifting* scheme for each source pixel:

$$\begin{bmatrix} l' \\ \alpha' \\ \beta' \end{bmatrix} = \begin{bmatrix} \frac{\sigma_t^l}{\sigma^l} \\ \frac{\sigma_t^\alpha}{\sigma^\alpha} \\ \frac{\sigma_t^\beta}{\sigma^\beta} \end{bmatrix} \circ \left(\begin{bmatrix} l_s \\ \alpha_s \\ \beta_s \end{bmatrix} - \begin{bmatrix} \mu_s^l \\ \mu_s^\alpha \\ \mu_s^\beta \end{bmatrix} \right) + \begin{bmatrix} \mu_t^l \\ \mu_t^\alpha \\ \mu_t^\beta \end{bmatrix} . \tag{4.12}$$

This method is very straightforward to implement and provides very convincing results as shown in Figure 4.7. However, the quality of the results depends on the composition similarity between images. For instance, an image with a large sky part and another one with a large ground part may produce undesirable results. To mitigate this issue, the computation of statistics has to be performed on a limited area, manually selected, with similar composition.

Although $l\alpha\beta$ are decorrelated for a natural image, they may not match with axes of principal components in some cases; e.g., urban photographs. Therefore, this can limit the effectiveness of results in such situations. A solution for these cases [145] is to compute principal components of I_s and I_t and to align the principal axes of I_s with the ones of I_t. Then, Equation 4.12 can be employed for the transfer.

Nevertheless, the RGB color space can still be employed without further color transformation when transferring mean values and covariance matrices for the source and the target images [298].

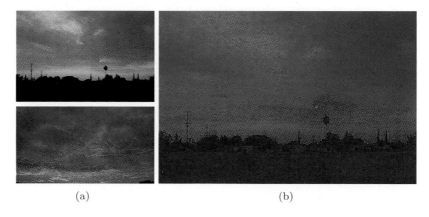

(a) (b)

Figure 4.8. An example of Monge–Kantorovicth matching applied to color transfer method [221]: (a) At the top, I_s; and at the bottom, I_t. (b) The result of the color transfer.

Pitié and Kokaram [221] established a general mathematical framework for linear transformation, proving that color transfer can be always be achieved when color distributions in I_s and I_t are both multivariate Gaussian distributions which are formally defined as

$$f_s(x) \propto \exp\left(-0.5(x - \mu_s)^T \mathbf{\Sigma}_s^{-1}(x - \mu_s)\right) \qquad (4.13)$$
$$f_t(x) \propto \exp\left(-0.5(x - \mu_t)^T \mathbf{\Sigma}_t^{-1}(x - \mu_t)\right),$$

where f_s and f_t are, respectively, the PDFs of I_s and I_t, and $\mathbf{\Sigma}_s$ $\mathbf{\Sigma}_t$ are, respectively, the covariance matrices of I_s and I_t. Given Equation 4.2.2, the framework defines the color mapping t such that $t(f_s(x)) \propto f_t(x)$ as

$$t(x) = \mathbf{T}(x - \mu_s) + \mu_t \qquad (4.14)$$
$$\mathbf{T}\Sigma_s \mathbf{T}^t = \Sigma_t. \qquad (4.15)$$

For example, Reinhard et al.'s method [229] for n color channel images can be defined as a diagonal matrix $\mathbf{T} = \mathrm{diag}(\sqrt{\sigma_t^1/\sigma_s^1}, ..., \sqrt{\sigma_t^n/\sigma_s^n})$. From this general formulation, novel transformations can be found. For example, the following mapping is leveraging on Monge–Kantorovicth theory:

$$\mathbf{T} = \Sigma_s^{-1/2}(\Sigma_s^{1/2}\Sigma_t\Sigma_s^{1/2})^{1/2}\Sigma_s^{-1/2}, \qquad (4.16)$$

which minimizes

$$\int_x \|t(x) - x\|^2 f_s(x)dx. \qquad (4.17)$$

Figure 4.8 shows the results of this novel mapping.

4.2.3   Gamut-Based Methods

An important issue, not always taken into account in color transfer methods, is to maintain retargeted colors in the source image, I_s, within the gamut of the target image, I_t, an problem seen previously in Chapter 3.

A first attempt in this direction [47, 48] clusters colors of both images into eleven color categories [272]. For each cluster and image, a convex hull (CH) is generated. Then, colors of the i-th corresponding category are matched as

$$I_o = \frac{\|c_s^i, I_s\|}{\|c_s^i, b_s^i\|} \left(b_t^i - c_t^i\right) + c_t^i, \qquad (4.18)$$

where c_s^i and c_s^i are, respectively, the centers of CH_s^i and CH_t^i, b_s^i is the intersection between CH_s^i and the ray from c_s^i to I_s, and b_t^i is the intersection between CH_t^i and the ray starting from c_t^i and same direction of the ray from c_s^i to I_s. This algorithm works well for paintings, but it has not been tested for more general categories of images such as photographs or computer generated images.

Similarly, automatic mood transfer [302] divides the color spectrum into 24 categories or moods by collecting images from the Internet; each category is generated from a set of 10 images, K. To change the *mood* of an input image, I_s, the user needs to select a mood keyword, w. Then, a mood histogram of I_s is generated based on the Euclidean distance in the RGB color space between I_s colors and each mood. To complete the transfer, w has to be the highest frequency in the histogram, and the second dominant color mood has to be 75% less than w. As proposed by Chang et al. [47], both colors of I_s and the closest mood's image, $I_t \in K$, are segmented into clusters; in this case, 24 clusters as the 24 moods. For each color c in an I_s cluster, the mapping is computed as the average between c and the corresponding centroid cluster of I_t.

Another gamut-friendly approach is to align the color gamut of I_s to the one of I_t [201]. To facilitate this task, the white points of both images are computed and normalized using a white balancing technique [94], which sets them to $(1, 1, 1)$ in the RGB color space. Note that this step reduces the difference between I_t and I_s. After white balancing, to further reduce computations, colors in both images are rotated from the $(1, 1, 1)$ axis to $(0, 0, 1)$; obtaining $\hat{I}_s$ and $\hat{I}_t$.

At this point, the luminance channel of $\hat{I}_s$ is matched to the one of $\hat{I}_t$ as:

$$\hat{L}_o = g(F_t^{-1}(F_s(L_s))), \qquad (4.19)$$

where F_s and F_t are, respectively, the CDFs of the histogram of L_s and L_t, and g is the gradient-preserving method [299] (see Section 4.2.1).

Then, gamut mapping is approximated by a linear transformation. Before this transformation, gamuts are shifted to share the same center as

$$I'_s = \hat{I}_s + \mu_s \quad \text{and} \quad \hat{I}'_t = I_t - \mu_s, \tag{4.20}$$

where μ_s and μ_t are, respectively, the mean values of $\hat{I}_s$ and $\hat{I}_t$. After this, $\mathbf{T}$, a rotation, is computed to put the gamut of I'_s inside the gamut of I'_s. $\mathbf{T}$ is defined as

$$\mathbf{T} = \begin{bmatrix} s_1\cos(\theta) & -s_1\sin(\theta) & 0 \\ s_2\sin(\theta) & s_2\cos(\theta) & 0 \\ 0 & 0 & 1 \end{bmatrix}, \tag{4.21}$$

where s_1 and s_2 are two scaling factors for chromaticities, and θ is the rotation around the luminance axis. These parameters are found by minimizing

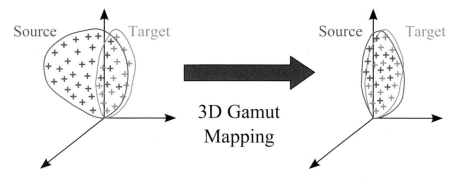

Figure 4.9. The main idea of Nguyen et al.'s color transfer method [201] is to align the gamut of I_s to the one of I_t.

a cost function, f, defined as

$$f(\mathbf{T}) = 2\mathcal{V}((\mathbf{T} \times CH_s) \oplus CH_t) - \mathcal{V}(\mathbf{T} \times CH_s), \tag{4.22}$$

where CH_s and CH_t are, respectively, the convex hull of I_s and I_t, $\mathcal{V}$ computes the volume of a convex hull, and $\oplus$ is the point concatenation operation between two convex hulls. Once parameters are found, the gamut can be rotated and shifted back as:

$$\hat{I}_o = \mathbf{T}\hat{I}'_s + \mu_t. \tag{4.23}$$

Finally, the output image, I_o, is obtained by rotating colors from the $(0,0,1)$ axis to the $(1,1,1)$ axis, and applying the white point of I_t; see Figure 4.10. This method was compared against the state-of-the-art showing that in most cases it ensures that retargeted colors are inside the gamut of I_t. Surprising, it was also shown that the use of RGB values with gamma encoding does not significantly affect the white balancing step and the other computations, when compared to linear RGB values.

(a)   (b)

Figure 4.10. An example of Nguyen et al.'s color transfer method [201]: (a) At the top, I_s; and at the bottom, I_t. (b) The result of the color transfer.

4.2.4 Local methods

Local methods for color transfer, as the name suggests, segment input images into soft or hard regions. Then, segments from I_t and I_s are typically matched, and a global color transfer method (previously seen) is applied.

A popular approach to segment I_t and I_s is to employ an expectation-maximization scheme [260, 296, 301].

In this scheme, the probability that a pixel color, c, belongs to the i-th Gaussian color distribution, $G_i(\mu_i, \sigma_i)$, is defined as

$$P(x) = \frac{\exp\left(-\frac{I(x) - \mu_i}{2\sigma_i^2}\right)}{\sum_{j=1}^{n} \exp\left(-\frac{c - \mu_j}{2\sigma_j^2}\right)}, \tag{4.24}$$

where I is an image, and μ_i and σ_i are estimated as

$$\mu_i = \frac{\sum_{x \in I} P(x) I(x)}{\sum_{x \in I} P(x)} \tag{4.25}$$

$$\sigma_i = \sqrt{\frac{\sum_{x \in I} P(x) \big(I(x) - \mu_i\big)^2}{\sum_{x \in I} P(x)}}, \tag{4.26}$$

where I is an image. The first step of these algorithms is to segment both I_s and I_t in the $l\alpha\beta$ color space by a set of n Gaussians. These Gaussians are initialized with k-means, then Equation 4.24 and Equation 4.26 are repeatedly applied until their values are stabilized, typically 3–5 iterations. Between Equation 4.24 and Equation 4.26, P is spatially smoothed using bilateral filtering; see Appendix A. Once Gaussians are computed for both I_s and I_t, a mapping between Gaussians in I_s and I_t is established by finding the closest in terms of mean and variance. The mapping between clusters can be further improved with specific policies for avoiding failures [117]. This can be achieved by firs classifying each cluster as a color style, with at least two different dominant colors, or a light style, with only one dominant color. Then for each different combination source and target clusters, in total four, a handling policy can be defined. For example, in the case of a source cluster has a light-style and the target cluster has a color style a photographic artistic approach can be taken into account [291]; the warmest colors of the source cluster will be mapped onto the highlights of the target cluster. Conversely cold colors will be mapped onto the shadows of the source cluster.

Finally, given a match, $\{G_{t_i}(\mu_{t_i}, \sigma_{t_i}); G_{s_j}(\mu_{s_j}, \sigma_{s_j})\}$ the transferring is simply achieved as

$$T(I_s(x)) = \sum_j P^j(x)\left(\frac{\sigma_{t_i}}{\sigma_{t_j}}\big(I_s(x) - \mu_{t_j}\big) + \mu_{t_i}\right), \qquad (4.27)$$

where P^j is the probability that $I_s(x)$ belongs to region t_j. These methods can be further extended on a general framework for soft color segmentation and other applications such as colorization, denoising, and deblurring [261]. Figure 4.11 shows results using local methods with and without matching policies.

A content-based approach segments each image based into semantic objects [293], then it applies a color transfer function to similar content. First, content in I_s and I_t is analyzed and extracted; different methods are employed for each type of content such as salient objects [87], people or faces [297], and background [114]. Figure 4.12 (a) shows a decomposition of a target and a source image. At this point, color transfer is directly applied to background parts [229] (see Section 4.2.2), e.g., the sky in I_s and I_t. This is also valid if there is only a salient object for each image. Otherwise, each sampled object is minimized using its centroid distance in obtaining matched pairs. Finally, each object in the pair is decomposed into superpixels [56] and matched with a soft mapping. This soft mapping is used to guide color transfer [229]; see Figure 4.12 (b).

4.2.5   User-Based Methods

In user-based methods, the user is required to select areas to be matched. Different tools can be used such as swatches, geometric selections, and strokes.

(a) I_s (b) I_t

(c) Tai et al. [260]   (d) Hristova et al. [117]

Figure 4.11. An example of a local method [260] with and without matching policies [117]. Note that the use of policies, (d), allows the transfer to be closer to the target, (b), than without (c). Original images are copyright of Fabrice Lamarche, and Hristina Hristova and Rémi Cozot helped generate them.

Swatches. Reinhard et al. [229] already pointed out that the main limitation of automatic methods is that I_s and I_t have to share similarity in their composition in order to obtain satisfactory results. To solve this issue, they proposed one of the first user-based solutions, which is swatch based. The user selects a pair of swatches from the source and the target, and the system computes statistics per swatch. Finally, the transfer is applied per swatch, where discontinuities are handled by blending colors based on the inverse

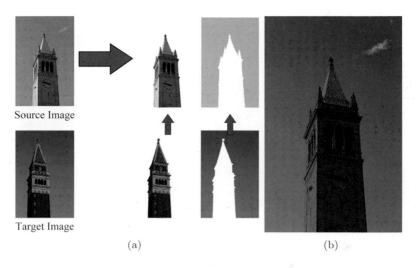

Source Image

Target Image

(a)   (b)

Figure 4.12. An example of content-based color transfer [293]: (a) At the top, I_s; and at the bottom, I_t. Note that salient objects and background are segmented into different layers and matched (red vertical arrows). (b) The result of the color transfer.

distance of the pixel from the center of a swatch, see Figure 4.13. The same or similar strategies are used in other works [288, 298]; see Section 3.10.

Maslennikova and Vezhnevets [182] proposed to compute weights for blending differently by computing a color influence map (CIM). Once the user has selected a region, R, from the target image, a mean, μ_t^R, and standard deviation, σ_t^R, are computed for this region only. Then, the CIM is computed as follows:

$$w(i) = g(\|I_s(i) - \mu_t^R\|_E) \quad g(x) = \exp(-3x^2), \qquad (4.28)$$

where i is the i-th pixel, and $\|\cdot\|_E$ is the Mahalanobis distance, which turns into the Euclidian distance in the $l\alpha\beta$ color space thanks to the decorrelated nature of this color space. At this point, Equation 4.12 is applied to the source, I_s, using μ_t^R and σ_t^R as target statistics and obtaining a temporary image, I_t. Finally, I_s and I_t are linearly blended using the CIM:

$$I_o(i) = I_s(i)(1 - w(i)) + I_t(i)w(i) \quad \forall i \in I_s. \qquad (4.29)$$

In Luan et al.'s system [167], the user first brushes a pair of swatches; one in the source image and the other in the target. Then, color transfer is applied using Equation 4.12. To avoid seams, an energy term is minimized similar to Poisson image editing [216] but without Dirichlet boundary conditions. The main advantage of this class is the fact that users have to just do a

(a)   (b)

Figure 4.13. An example of the swatch based method [229]: (a) At the top, I_s; and at the bottom, I_t. Note that matched swatches have the same color. (b) The result of the color transfer.

simple selection—a swatch. However, shape selection may be not optimal, e.g. a square [182, 229], for including important color features or it could involve a lot of manual work [167].

Strokes. To overcome the pitfalls of swatch-based methods, researchers have introduced stroke-based methods, which are very popular in editing images [8, 156, 161]. In this class of algorithms, users quickly specify, by drawing a few rough strokes, editing parameters at a couple of sparse locations. The system automatically propagates these editing parameters from these sparse locations to the whole image.

Chen et al. [289] proposed a system with a stroke-based interface. In this system, the user draws pairs of strokes in corresponding regions in the source and the target image to specify which regions need to be transferred from the target to the source. Regions to be transferred are called foreground, $\mathcal{F}$, others are called background, $\mathcal{B}$; see Figure 4.14 (a).

After this step, the system segments the source image into foreground and background, generating two pixel point sets, $P_{\mathcal{F}}$ and $P_{\mathcal{B}}$. This is achieved by performing a modified graph-cuts version [37] that works in the CIE $L^*a^*b^*$ color space and takes into account distances from strokes. Once each pixel has a foreground or background label, Equation 4.12 is generalized taking into account the segmentation and different strokes, obtaining

$$I_o(i) = (I_s(i) - u(i))f(i) + v(i) \quad \forall i \in I_s, \quad (4.30)$$

where u, f and v are strokes-aware and spatially varying values for the

(a)   (b)

Figure 4.14. An example of a stroke-based method [289]: (a) At the top, I_s with foreground (red) and background (green) strokes; and at the bottom, I_t with strokes matching the ones in I_s. (b) The result of the color transfer.

shifting, scaling, and then shifting again functions. These are defined as

$$u(i) = \begin{cases} \sum_{j \in P_{\mathcal{F}}} C(I_s(i), j)\mu_s^j + \sum_{j \in P_{\mathcal{B}}} C(I_s(i), j)I_s(i) & \text{if } l(i) = \mathcal{F} \\ I_s(i) & \text{otherwise} \end{cases}$$
(4.31)

$$f(i) = \begin{cases} \sum_{j \in P_{\mathcal{F}}} C(I_s(i), j)\frac{\sigma_t^j}{\sigma_s^j} + \sum_{j \in P_{\mathcal{B}}} C(I_s(i), j) & \text{if } l(i) = \mathcal{F} \\ 1 & \text{otherwise} \end{cases}$$
(4.32)

$$v(i) = \begin{cases} \sum_{j \in P_{\mathcal{F}}} C(I_s(i), j)\mu_t^j + \sum_{j \in P_{\mathcal{B}}} C(I_s(i), j)I_s(i) & \text{if } l(i) = \mathcal{F} \\ I_s(i) & \text{otherwise} \end{cases}$$
(4.33)

where $G_s^j(\mu_s^j, \sigma_s^j)$ and $G_t^j(\mu_t^j, \sigma_t^j)$ are Gaussian color models for each j-th corresponding foreground stroke, $P_{\mathcal{F}}$, and background ones, $P_{\mathcal{B}}$. $C(I_s(i), j)$ determines how the $I_s(i)$ color has to be influenced by the j-th stroke and is defined as

$$C(I_s(i), j) = \frac{P(I_s(i)|G_s^j)}{\sum_k P(I_s(i)|G_s^k)},$$
(4.34)

where $P(I_s(i)|G_s^j)$ is a Gaussian PDF estimating the belongingness of a color, $I_s(i)$, to the color model G_s^j of the j-th stroke in I_s. In order to mitigate artifacts in the pixel-wise color transfer, Equation 4.30, u, v, and f are filtered using an edge-stopping weighted least square optimization [161]. The main advantage of the proposed system is the rapid learning curve; non-experts in photographic retouching can easily obtain satisfactory results;

see Figure 4.14 (b). However, the system can have issues when transferring details in strongly overexposed and underexposed areas.

Dong and Xu [71] extended the CIM approach by Maslennikova and Vezhnevets [182] replacing swatches with strokes. In their system, the user draws a few strokes for defining the foreground (to transfer or to be transferred) and background labels in the source and the target images. Then, labels are propagated using GrowCut [274] and during this process Equation 4.28 and are applied. Note that they proposed a slightly more general $g(x)$; $g(x) = \exp(-x^2/\lambda^2)$ where $\lambda \in [0, 1]$ is a parameter which controls color propagation strength.

An and Pellacini [9] proposed a stroke-based system where the transfer functions are propagated to the whole image. These functions are defined using a parametric transfer model in any color space where luminance and chrominance values are separated, such as CIE $L^*a^*b^*$ or oRGB [38]. This is formally defined as

$$F_p : c \to c', \qquad (4.35)$$

where $c = (l, a, b)$ is the input triplet (l is the luminance value and a and b are the chrominance values), and $c' = (l', a', b')$ is the transformed triplet. F_p is modeled differently for each component. For the luminance values, a piecewise cubic Bézier spline, $\hat{l}$, in the plane (l, l') is employed, while in the chromaticity domain, (a, b), an affine transformation cubically interpolated along l is used and defined as

$$[a', b', 1]^T = \hat{M}(u)[a, b, 1]^T \quad \text{where } u = \hat{l}^{-1}(l) \qquad (4.36)$$

where $\hat{M}$ is the interpolated affine matrix. To avoid luminance clipping or flipping of luminance gradients, three constraints are imposed:

- $F_p(0, a, b) = (0, a', b')$;

- $F_p(1, a, b) = (1, a', b')$;

- $\hat{l}$ is monotonically increasing.

To take into account the fact that CIE $L^*a^*b^*$ chromaticity values close to black or white are noisy, the control point at $l = 0$ is the same as the next, and that the one at $l = 1$ is the same as the previous control point. Moreover, C^1 continuity is imposed. Given a pair of strokes, a source and a target stroke, F_p parameters are computed by minimizing the differences between their color distributions. This can be expressed as

$$\arg \min_{\lambda} E(\lambda) = \int \|F_s(x) - F_t(x)\|^2 dx, \qquad (4.37)$$

where F_s and F_t are, respectively, the three-dimensional cumulative density function of source and target strokes in the CIE $L^*a^*b^*$ color space.

Equation 4.37 is solved using the simplex algorithm where the $\hat{l}$ is initialized as the identity and the affine matrix is initialized using a Gaussian model [229]. Once the model parameters are estimated for each pair of strokes, parameters are computed for all pixels by adopting edit propagation methods [8, 161], and then the model is applied per pixel. This allows us to maintain the rich details found in the original image and avoid artifacts such as loss of contrast and texture details [8]. This approach usually reduces trial-and-error time for doing adjustments, and subtle details can be successfully transferred, especially for skin appearance [9].

Sparse Correspondences. A different approach for user-based methods is to use sparse correspondences. In this case, the user typically defines a color mapping by clicking correspondences in the source and the target images. For example, Oskam et al. [206] proposed a sparse correspondence color

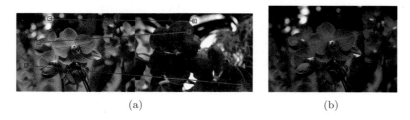

(a)   (b)

Figure 4.15. An example of the Oskam et al. [206] framework: (a) An implementation showing picked correspondences. (b) Final transfer results; note that the transfer is successful, but the method presents similar global methods' limitations; e.g. closed flowers have a mixed color between violet and green because of the lack of the use of spatial information.

editing framework in the CIE $L^*a^*b^*$ color space with many applications such as color transfer, color correction, and augmented reality color balancing. In their system, when used for color transfer, the user creates a set of pairs in the target and the source image. For each pair of colors, (p_s^i, p_t^i), a p_s^i is defined as the *support* and their difference vector $\mathbf{v}^i = \|p_t^i - p_s^i\|$ as the *data value*. Each color, e_s, in the source image is re-mapped by adding to it a translation vector, $\mathbf{v}(e_s)$, which is computed as

$$\mathbf{v}(e_s) = \frac{1}{\sum_{i=1}^{n} \phi_i(\|e_s - c_s^i\|)} \sum_{i=1}^{n} \phi_i(\|e_s - c_s^i\|)\mathbf{w}^i, \qquad (4.38)$$

where $\phi(x) = (1 + x)^{-\epsilon}$ with $\epsilon = 3.7975$ (determined after experiments), and weights, $\mathbf{w}^i$, are computed through minimization:

$$\arg\min_{\mathbf{w}^i} \left(\|\mathbf{v}^i - \mathbf{v}_i(c_s^i)\|^2 \right). \qquad (4.39)$$

Note that this leads to a linear system, $v = \Phi w$, for each color channel. Results shows this method can match Reinhard et al.'s work [229]. However, it suffers from similar limitations, because the vector field has a global nature; see Figure 4.15 (b). Actually, spatial information is not used for computing $\mathbf{v}^i$.

Recently, researchers have proposed more general stroke-based editing systems [80, 192] for solving the edit propagation problem [8]. These systems also allow users to achieve color transfer, but they require more user inputs than specific methods seen before.

4.2.6   Color Consistency in Collections

Color consistency in collections is the problem of enforcing consistent color appearance of shared content across multiple photographs. This is an important problem in image editing and computer vision. For example, computer vision applications improve in results quality when appearance is consistent for many tasks such as features extraction and matching, structure-from-motion, stereo matching or multi-view stereo, panorama stitching, etc. Furthermore, people nowadays tend to collect a huge number of images when they travel for leisure or work, during meals, and in general, in every event of their lives. However, these images are usually captured using smartphones or tablets, which typically provide an automatic mode for different settings such as color balance, exposure, ISO, etc. Therefore, images may have an inconsistent look from one photograph to another, and correcting such a large collection may require a lot of time when done manually.

Exploiting content. Given the content redundancy, a typical approach is to find correspondences between shared content in images, and then to compute a color mapping f. Figure 4.17 shows an example of these methods.

In color consistency, correspondences between pixels can be computed using different methods such as sparse methods which are based on local features [40, 83, 135, 205] (e.g., SIFT [166]), and dense methods, which can be based on disparity [110, 247], optical flow [281], SIFT-Flow [163], and extended generalized PatchMatch [106, 107]. While sparse methods create few matches; see Figure 4.16 (c), dense methods output many matches; see Figure 4.17 (c). Therefore, these last methods are typically more robust and can produce more convincing results in challenging cases.

Once correspondences between images are found, the color mapping, f, needs to be computed. Typically, this is done by minimizing an energy term, E, which can be defined as

$$E = \sum_{i=1}^{n} \bigl(f(I_s(m_s^i)) - I_t(m_t^i)\bigr)^2, \tag{4.40}$$

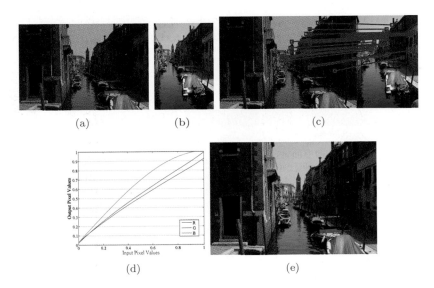

(a)					(b)					(c)

(d)					(e)

Figure 4.16. An example of a SIFT-based method for color consistency [83]: (a) A source image. (b) A target image. (c) SIFT matches. (d) Color transform curve for each color channel. (e) The image in (a) after applying the color transform (d); note that colors are matched with the ones in (b).

where I_s and I_t are, respectively, the source and the target image, (m_s^i, m_t^i) are the coordinates of matched pixels, and n is the number of matched pixels. f can vary a lot; it can be a cubic spline for each color channel [84, 106], a color 3×3 matrix [247], a gain factor [40], and gamma and gain [110]. Moreover, any color transfer method seen before (e.g., Reinhard et al.'s method; Section 4.2.2) in this chapter can be also employed, as in the case of Oliveira et al. [205] based on CIM; see Section 4.2.5.

When there are more than two photographs, i.e. a collection, a *match graph* helps in reducing computations while minimizing Equation 4.40. A match graph [107] is a graph $G(V, E)$ where vertices, V, are photographs in the collection, and edges, $E_{ij} = (V_i, V_j)$, provide correspondence information between two photographs. This graph is computationally expensive to fully compute, because correspondences need to be found for a quadratic number of pairs. HaCohen et al. [107] proposed a learning approach, i.e., support vector machines [57], for predicting links. The trained SVM (using full graphs for learning) can determine the likelihood that a link between an image pair (I_i, I_j) exists or not in very short time. For example, the match graph dataset of 865 images can be computed in less than 7 hours instead of a week.

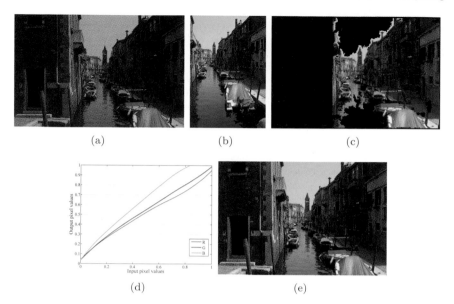

(a)   (b)   (c)

(d)   (e)

Figure 4.17. An example of HaCohen et al.'s method [106]: (a) A source image.
(b) A target image. (c) Found dense matches between (a) and (b). (d) The color
transform; a spline. (e) The image in (a) after applying the color transform (d);
note that colors are matched with the ones in (b).

Histogram-based approaches. Histograms can also be exploited in large
collections; these are computationally cheap to compute and manipulate.
Moreover, histograms of such photographs, i.e., the same scene, can exhibit
similar color distributions.

A straightforward method is to align histograms of photographs for each
color channel [245]. The alignment of histograms requires extraction of
features for each histogram; which are typically local maxima computed in
scale space [160] to reduce outliers and noise influence. Given two gray-scale
images, I_1 and I_2, their respective histograms are computed, h_1 and h_2.
Then, feature vectors are extracted, $\vec{f}_1$ and $\vec{v}_2$, and matched using the
minimum Euclidean distance. After this step, the color mapping, f, is a
polynomial whose coefficients are found by minimizing an energy term, E,
defined as

$$E = \sum_{i=1}^{n} \|f_i(m_1(i)) - m_2(i)\| \qquad f(x) = \sum_{k=0}^{p} c_k x^k, \qquad (4.41)$$

where the couple $\big(m_1(j); m_2(j)\big)$ is a match, and n is the number of matches.
Equation 4.41 leads to a linear system. For color images, this process is
repeated for each color channel separately; it can be also extended to

multiple images by modifying Equation 4.41. The method outperforms moments transferring under different lighting conditions, scale, and small shadow changes [245].

A more general approach, proposed by Papadakis et al. [209], is a variational formulation for midway histogram equalization (MHE) [67]. Given two gray-scale images, I_1 and I_2, and their cumulative histograms, F_1 and F_2, MHE is defined as

$$\hat{F} = \left(\frac{F_1^{-1} + F_2^{-1}+}{2} \right)^{-1},$$ (4.42)

where $\hat{F}$ is a midway cumulative histogram, which can be employed for remapping both I_1 and I_2 to share it. The variational formulation extends MHE to color images and multiple images by defining an energy, E, to minimize, which respectively converts I_1 and I_2 into $\hat{I}_1$ and $\hat{I}_2$, which share a common intermediate 3D histogram as intermediately as possible to their original ones. The main building term of this energy is defined as

$$E(\hat{I}_1, \hat{I}_2) = \frac{1}{2} \int_0^1 \int_0^1 \int_0^1 \left(\mathbf{F}_1(\vec{x}) - \mathbf{F}_2(\vec{x}) \right)^2 d\vec{x}+$$ (4.43)

$$\frac{1}{2} \int_0^1 \int_0^1 \int_0^1 \left(\hat{\mathbf{F}}_1(\vec{x}) - \hat{\mathbf{F}}_2(\vec{x}) \right)^2 d\vec{x},$$

where $\mathbf{F}_1$ and $\mathbf{F}_2$ are, respectively, the cumulative 3D histograms of I_1 and I_2, and $\hat{\mathbf{F}}_1$ and $\hat{\mathbf{F}}_2$ are, respectively, the cumulative 3D histograms of $\hat{I}_1$ and $\hat{I}_2$. Two conservative terms are added to E for maintaining unaltered colors, and image geometry; i.e., gradients. Note that E can be simply extended to handle datasets of n images as

$$\arg \min_{\hat{I}_1,...,\hat{I}_n} \sum_{1 \leq k < l \leq n} E(\hat{I}_k, \hat{I}_l).$$ (4.44)

This method can produce convincing results (see Figure 4.19) and it can handle dark images, which are the most difficult kind of images due to possible noise magnification. Furthermore, the method can also applied to color transfer (Section 4.2) with success; see Figure 4.18.

4.2.7   Color Harmonization

Color harmonization is the process by which colors of a source image, I_s, become harmonic either with respect to themselves or to colors of a target image I_t. What is a harmonic color? Harmonic colors have a special internal relationship that provides a pleasant visual perception. This harmony is

(a)   (b)

Figure 4.18. An example of color transfer using the variation model for histogram matching [209]: (a) The target image. (b) The source image. (c) The result of color transfer. Nicolas Papadakis helped in the generation of these images.

Figure 4.19. An example of consistency in collections using the variation model for histogram matching [209]: In the top row, the input sequence of images. In the bottom row, the sequence in the top row after application of photo consistency. Nicolas Papadakis helped in the generation of these images.

due to the relative positions of colors in color space. Amongst artists, there is no precise definition of harmony, but only schemes and relations in a color space that can help when choosing a color set [183, 267]. Furthermore, color patterns can be extracted from images [202]. In computer graphics, the problem of achieving color harmonization was first introduced by Cohen-Or

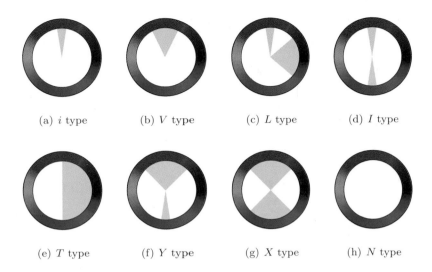

(a) i type   (b) V type   (c) L type   (d) I type

(e) T type   (f) Y type   (g) X type   (h) N type

Figure 4.20. The eight color templates on the hue wheel. Colors in the gray sectors are typically called *harmonic* colors. The large sectors in V, Y, and X have a central angle of 93.6°. The small sectors in i, L, I, and Y have a central angle of 18°. The large sector in L has a central angle of 93.6°, and the large sector in T has a central angle of 180°. The N type corresponds to gray-scale images. Note that these templates can be rotated by an arbitrary angle; $\alpha \in [0, 2\pi]$.

et al. [55]. In their seminal work, they distilled the harmony definition and some schemes developed by Matsuda [183] and Tokumaru [267], which were derived by Itten [126]. This led to the eight harmonic types, $T_{m(\alpha)}$ ($m \in \{i, I, L, T, V, X, Y, N\}$), which can be rotated by an arbitrary angle $\alpha \in [0, 2\pi]$; these templates are shown in Figure 4.20. These are defined in the hue color channel of the HSV color wheel: colors that are inside the gray areas are meant to be harmonic. Moreover, they defined a two-stage approach for color harmonization:

- **Template estimation** determines $T_{m(\alpha)}$ from I_s or I_t; for *harmonization transfer*. This can be also manually defined using a simple interface.

- **Color harmonization** applies the found template to I_s.

Template Estimation. Given these templates, the sector border hue of T_m that is closest to the hue of pixel p, is defined as $E_{T_m(\alpha)}(p)$.

At this point, it is possible to define a function that measures the distance between the hue of an image and a template as

$$F(I, (m, \alpha)) = \sum_{p \in I} \|H(p) - E_{T_m(\alpha)}\| \cdot S(p), \qquad (4.45)$$

where H and S are, respectively, the hue and saturation channels, and $\| \cdot \|$ is the arc-length distance on the hue wheel (in radians), which is zero if $H(p)$ is inside the sector T_m. Note that the use of S is to cope with low saturation colors, which are perceptually less noticeable by the HVS.

Exploiting Equation 4.45, T_m and α can be estimated for an input image I through minimizing Equation 4.45 as

$$B(I) = (m_0, \alpha_0) \quad \text{such that} \quad (m_0, \alpha_0) = \arg\min_{m, \alpha} F(I, m, \alpha). \qquad (4.46)$$

Brent's algorithm [225] can be used for this minimization. When I_s has to be harmonized with respect to I_t, $B(I_t)$ has to be computed, and not $B(I_s)$.

One issue of the cost function F in Equation 4.45 is the loss of meaningfully minor colors, and the possibility for abrupt and unnatural changes for skin or sky colors. This can be avoided by using a saliency model [155] as proposed by Baveye et al. [28]. In this case, two templates are computed: one, (m_0, α_0), for the source image or the target one (depending on the application); and another, (m_S, α_S), for only the salient pixels in the source image. A template, (m, α), can define a distribution, D, as

$$D(p, m, \alpha) = \begin{cases} \sum_{k=1}^{n} DS(p, k) & \text{if } \sum_{k=1}^{n} DS(p, k) \neq 0 \\ \frac{0.01}{360} \cdot \sum_{p \in I} \left(\sum_{k=1}^{n} DS(p, k) \right) & \text{otherwise} \end{cases}$$

$$(4.47)$$

where n is the number of sectors in $T_{m(\alpha)}$, and DS is the distribution sector, which is defined as

$$DS(p, k) = \begin{cases} \exp\left(-\left(1 - \left(\frac{2\|H(p) - C_k\|}{w_k}\right)^{10}\right)^{-1}\right) & \text{if } p \in \left(C_k - \frac{w_k}{2}, C_k + \frac{w_k}{2}\right) \\ 0 & \text{otherwise} \end{cases}$$

$$(4.48)$$

where C_k and w_k are, respectively, the center and size of the k-th sector in a template. After computing the distributions for (m_0, α_0) and (m_S, α_S), these are merged as

$$D'(p) = \max\left(D(p, m_0, \alpha_0), D(p, m_S, \alpha_S) \right), \qquad (4.49)$$

from which a new template is extracted. This whole process automatically enforces preservation of colors in regions determined by a saliency map.

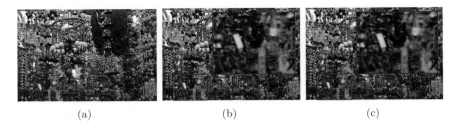

(a)   (b)   (c)

Figure 4.21. An example of color harmonization labeling [55]: (a) The original source image. (b) A harmonized version of (a) with a naïve approach for labeling; note in the zoomed green areas the discontinuities, which lead to artifacts. (c) A harmonized version of (a) using graph-cut; note that artifacts are not present.

Color Harmonization. Once a harmonic scheme, $T_m(\alpha)$, is computed from an input image I_t (see Equation 4.46) or manually defined, hue values of pixels can be shifted toward the defined sector in $E_{T_m(\alpha)}$; this process is called *color harmonization*. As a first step, each sector of $T_m(\alpha)$ has to be assigned to a pixel p, if the template has more sectors. A naïve approach is to assign p to the nearest sector, in terms of hue, but this does not take into account spatial coherency, which leads to artifacts; see Figure 4.21 (b). To avoid this issue for templates with two sectors, a graph-cut approach is employed [37] to determine the labeling $V = \{v(p_1), v(p_2), ..., v(|\Omega|)\}$ for each pixel p. More formally, this can be defined as an energy, $E(V)$, to minimize:

$$E(V) = \lambda E_1(V) + E_2(V), \qquad (4.50)$$

where $E_1(V)$ takes into account the distances between the house $H(p)$ and $H(v(p))$, and it is defined as

$$E_1(V) = \sum_{i=1}^{|\Omega|} \|H(p_i) - H(v(p_i))\| \cdot S(p_i). \qquad (4.51)$$

$E_2(V)$ promotes color coherence between neighboring pixels assigned to the same label, and it is defined as

$$E_2(V) = \sum_{\{p,q\} \in N} \frac{\delta(v(p), v(q)) \cdot \max\{S(p), S(q)\}}{\|H(p) - H(q)\|}, \qquad (4.52)$$

where N is the set of neighboring pixels in Ω, and δ is defined as

$$\delta(x, y) = \begin{cases} 1 & \text{if } x \neq y \\ 0 & \text{otherwise} \end{cases}. \qquad (4.53)$$

Figure 4.21 (c) shows a result of applying such a technique. Graph-cut can fail in some cases, but a two-level graph-cut (i.e., region and pixel levels) can reduce these cases [263]. However, graph-cut is a computationally expensive method; especially for large or multiple images. A reasonable trade-off between speed and quality is histogram splitting [242], or a k-means clustering variant, ACoPa [68], can provide satisfactory results [28].

After solving Equation 4.50, the hue of each pixel $p \in I$ has to be shifted toward the associated $E_{T_m(\alpha)}(p)$. This can be achieved by using the following formula:

$$H'(p) = H_C(p) + \frac{w}{2}\left(1 - \exp\left(-\frac{\|H(p) - H_C(p)\|^2}{2\sigma^2}\right)\right) \qquad \sigma = \frac{w}{2}, \ (4.54)$$

where $H_C(p)$ is the central hue of the sector associated with the pixel p, and w is the arc-width of the template. Note that this process shares similarities with gamut mapping; see Section 2.3. Figure 4.22 shows an example of color harmonization of a source image I_s with respect to a target image I_t. In this case, $B(I_t) = (m_t, \alpha_t)$ was computed and used as input in Equation 4.50 and Equation 4.54.

Note that color harmonization can also be treated as a reassigning process [262] which allows skipping the labeling task or segmentation of the hue histogram. In this case, an energy function, $E(H)$, needs to be minimized to obtain the harmonized H channel. This energy can be defined as

$$E(H) = \sum_{p \in I}\left(H(p) - \sum_{q \in N(p)} w(p,q)H(q)\right)^2 + \lambda \sum_{p \in I} g(H(p), T_{m(\alpha)}), \ (4.55)$$

where g is a function enforcing hue shifting toward the template sectors, λ determines g enforcement (a large number for enforcing), and w is a weight function defined as [156]

$$w(x,y) = \frac{\exp\left(-\frac{\left(H(x) - H(y)\right)^2}{2\sigma_x^2}\right)}{\sum_{t \in N(x)} \exp\left(-\frac{\left(H(x) - H(t)\right)^2}{2\sigma_x^2}\right)}, \qquad (4.56)$$

where σ_x is the deviation of the hue values in a window around r. The minimization can be solved using the conjugate gradient algorithm [225].

Data-Driven Harmonization. Cohen-Or et al.'s framework [55] is not the only possibility for harmonizing images given a color palette. Another popular approach is the data-driven one, where the user typically chooses from a database of color patterns [277] or a word with an associated hue

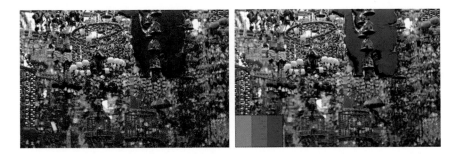

Figure 4.22. An example of color harmonization [55]: On the left side, the original source image. On the right side, the source image after harmonization with respect to the pattern in the bottom-left corner.

histogram [115]. Then, this pattern is automatically applied to an input image that needs to be harmonized.

In color conceptualization [115], a large number of images are classified in a database based on their semantic content such as *coast, open country, forest*, etc. Then, images for each label are clustered based on the hue histogram, and each cluster is manually labeled with a subjective description such as *cold* or *warm*. To harmonize an input image, I, with respect to the selected word, w, the hue peaks in the hue histogram of I, are modified to correspond to the ones in the hue histogram associated with w. Note that, given w, the system chooses the cluster of w with the closest hue histogram to the one of I. Another approach, by Wang et al. [277], automatically decomposes the image into $K \in [5, 12]$ soft segments [8]. The user has to select a color pattern in a database, and then the system recolors the K segments with an optimization that balances colors in the pattern and the original colors. To achieve this, for each segment a texture descriptor is extracted and matched with the others in a database; each entry in this database has an associated color PDF that is used as a constraint in the optimization.

However, these methods give little control to the user. To solve this issue, Chang et al. [46] proposed a user interface (UI) that allows users to modify the input image palette. Once an image is loaded in the system, its palette is computed [202], and the user can modify each color in the palette with a color wheel. Then these changes are propagated in the whole image using radial the basis function

$$f(x) = \sum_{i=1}^{k} w_i(k) f_i(x) \quad w_i(x) = \sum_{j=1}^{k} \lambda_{ij} \exp\left(-\frac{\|x - C_j\|^2}{2\sigma_r^2}\right), \quad (4.57)$$

such that

$$\sum_{i=1}^{k} w_i(x) = 1, \tag{4.58}$$

where C_i is the i-th color in the original palette, σ_r is the mean distance between all pair colors in the palette, λ_{ij} are coefficients computed by solving a linear system where $w_i(C_i) = 1$ and $w_{j \neq i} = 0$, and $f_i(x) = x'$ is defined as

$$\frac{\|x' - x\|}{\|C_i' - C_i\|} = \min\left(\frac{\hat{x} - x}{\hat{C}_i - C_i}\right), \tag{4.59}$$

where C_i' is the new palette color, $\hat{C}_i$ is the intersection of the ray from C_i to C_i' with the gamut boundary, and $\hat{x}_i$ is the intersection of a parallel ray with the boundary starting from x_i.

(a)   (b)

Figure 4.23. An example of the palette interface by Chang et al. [46]: (a) At the top, I_s; and at the bottom, I_t. Note that below each image there is its palette. (b) The result of the color transfer after matching the source palette with the target one. Huiwen Chang helped in the generation of this result.

4.3   Style Transfer

Style transfer between images is the general problem of transferring from I_t onto I_s, not only colors, but also contrast, details, etc. All these features define more generally the style of a image or its photographic look. Therefore, it is also a more complex problem.

Image analogies [113] are strongly related to style transfer. In fact, the style of an edited image, I'_t, is transferred onto a novel one, I_t. A coarse-to-fine (i.e., Laplacian pyramids) algorithm is employed for computing, at each level, the best matches (neighborhood matching) between pixels I'_t and I_s exploiting the knowledge of the original image I_t. Each best match value is then directly copied from I'_t to the query pixel of the output image at the current working level. Although this method can generate good results, a target pair is required as input: the original image and a version of it with a style. Therefore, this algorithm solves a more specific problem.

4.3.1 General Content Style Transfer

A seminal work in style transfer, proposed by Bae et al. [21], is based on a two-scale nonlinear decomposition of the image achieved using the bilateral filter; see Appendix A. As a first step, the method decomposes each image, I, into a base, B, and a detail, D, layer as

$$B = BF(I', \sigma_s, \sigma_r) \quad \text{and} \quad D = I - B, \qquad (4.60)$$

where $I' = \log(I)$, $\sigma_s = \min(\text{width}, \text{height})/16$ and $\sigma_r = p_{90}(\|\nabla I'\|)$ (p_x denotes the x-th percentile). This decomposition can suffer from artifacts; i.e., gradient reversal [74]. Therefore, gradients are forced in D to have the same sign and similar magnitude of the input derivatives in I' by using Poisson image editing [216]. Once D is fixed, B is subsequently updated.

At this point, large-scale spatial distribution in B_t are transferred onto B_s using histogram matching obtaining, B'_s; see Section 4.2.1.

(a) (b) (c)

Figure 4.24. An example of the two-scale decomposition's method [21] for style transfer: (a) The input source image. (b) The input target image with a *gritty* style. (c) The result of the style transfer method. Note that the style is successfully transferred from (b) to (a).

Then, there is the need to transfer textureness from the target to the source image. In order to achieve this, the amount of texture in an image is analyzed to determine regions with high levels of detail. Similar to previous work [157], an activity map can be employed and computed as

$$H = |I - (I * G_{\sigma_s})| * G_{\sigma_s}, \qquad (4.61)$$

where σ_s is the same as in Equation 4.60. However, this activity map can lead to artifacts at boundaries between high- and low-frequency regions. To solve this issue, the activity map is filtered using a cross-bilateral filter as

$$T(I) = CBF(H, I, 8 \cdot \sigma_s, \sigma_r). \qquad (4.62)$$

Once the activity map is defined, texture details are transferred by applying histogram matching from $T(\log(I_t))$ to $T(\log(I_s))$ to obtain the T'. However, it has to be scaled to avoid halos as

$$\rho = \max\left(\frac{T' - T(B'_s)}{T(D_s)}, 0\right). \qquad (4.63)$$

Finally, B'_s and D_s layers are recombined as

$$O_s = \exp(B'_s + \rho D_s). \qquad (4.64)$$

Note that Equation 4.64 can result in saturated highlights and shadows. These are avoided by applying histogram matching between O_s and I_t, and adding constraints on the gradient field to preserve details as in the original I_s.

The method can achieve convincing style transfer, see Figure 4.24, but it can suffer from noise and compression artifacts, which can be boosted after the application of the algorithm. These issues are more apparent on human faces; a straightforward solution is to extract background and foreground computing or drawing a mask. Then, two transfers, one for each component, are independently applied [35]. This step has to be followed by edge-persevering filtering to avoid artifacts around edges of the mask.

Style transfer can be also achieved using a multi-resolution pyramid representation without downsampling and Haar filters [258]. This may also mitigate exaggerated artifacts. The key contribution is a smoothed histogram matching technique. A naïve histogram matching can distort the shape of subbands, and this results in haloing, too strong edges, and the amplification of noise and blocking artifacts. The distortion of the k-th subband can be quantified as

$$g_k(B_s^k(i), B_t^k(i)) = \log(|B_o^k(i)|) - \log(|B_s^k(i)|) \qquad (4.65)$$
$$B_o^k(i) = \text{histmatch}(B_s^k(i), B_t^k(i)),$$

where i is the pixel index, and B_s^k and B_t^k are, respectively, the k-th subbands of I_s and I_t. Note that while $g_i > 0$ means there is an increase in

(a)   (b)

Figure 4.25. A comparison between style transfer methods using the source image, Figure 4.24 (a), and the target image, Figure 4.24 (b): (a) The two-scale decomposition method [21]. (b) The multi-resolution's method [258].

the coefficient magnitude; i.e. details are enhanced, $g_i < 0$ vice versa; i.e., details are dampened. To minimize artifacts, large values in $g_k(i)$ need to be avoided, which can be achieved by controlling the maximum gain as

$$\hat{g}_k(B_s^k(i), B_t^k(i)) = \exp\left(\frac{\delta_k}{g_{k,\max}}g_k(B_s^k(i), B_t^k(i))\right), \qquad (4.66)$$

where $g_{k,\max}$ is the maximum g_k value, and $\delta_k = 1.5$ indicates the maximum allowed gain for the subbing at the k-th level. As proposed by Li et al. [157], the new gain is not directly applied to B_s^k, but two activity maps are computed by convolving subbands as

$$A_s^k = N(\sigma) * |B_s^k| \qquad A_t^k = N(\sigma) \circ |B_s^k|, \qquad (4.67)$$

where N is a Gaussian kernel with σ increasing by a factor two between levels with the value at the finest scale set to four. These activity maps are spatially smooth, reducing distortion. Then, the transfer is computed as:

$$B_h^k(i) = m_k \hat{g}_k(B_s^k(i), B_t^k(i)) \times B_s^k(i), \qquad (4.68)$$

where m_k is a scaling factor related to the subband; this linearly reduces from 1.0 at the finest scale to 0.45 at the coarsest scale. Although the regularization in Equation 4.68 reduces most of the artifacts, strong edges may be amplified in some cases when doing such manipulation. Therefore, B_h^k and B_s^k are blended using the aggregate activity map as a weight; this is defined as

$$A_s^+ = f_C(\sum_{k=0}^{m} A_s^k), \qquad (4.69)$$

where f_C clamps its input to 0 at the 85th percentile and 1 at the 95th percentile. A result of the proposed method is shown; Figure 4.25 (b). As it can be seen, the method successfully transfers color and style without artifacts. Moreover, it can be used in a more general framework for other applications such as high-quality seamless cloning; noise can be matched using a similar multi-resolution scheme.

A more general and mathematical-based method for reducing artifacts has been presented by Rabin et al. [226, 227]. In their work, they introduce a transportation map defined as

$$M(I_s) = T(I_s, I_t) - I_s. \tag{4.70}$$

M can be regularized by using Yaroslavsky filters [304] as

$$[Y_{I_s} M(I_s)] = \frac{1}{C} \int_{y \in N(x)} [M(I_s)](y) e^{-\frac{\|I_s(x) - I_s(y)\|^2}{2\sigma^2}} dy, \tag{4.71}$$

where $N(x)$ is a spatial neighborhood of the x-th pixel (with a radius of 10 pixels), $\sigma = 10$, and C is a normalization term defined as:

$$C = \int_{y \in N(x)} e^{-\frac{\|I_s(x) - I_s(y)\|^2}{2\sigma^2}} dy. \tag{4.72}$$

Then, the final regularization image, Transportation Map Regularization (TMR), is defined as:

$$\text{TMR}(T(I_s)) = I_s + [Y_{I_s} M(I_s)] \tag{4.73}$$

$$= Y_{I_s}(T(I_s)) + \left(I_s - Y_{I_s}(I_s) \right). \tag{4.74}$$

Note that $T(I_s)$ is basically filtered by a non-local means operator. More iterations of the process are typically required to be effective in practice. Since this process numerically converges to a stable value, it is applied until the error between the previous and next TMR is less than a given threshold t ($t = 1$ for 8-bit images); see Figure 4.26.

Data-Driven Style Transfer. In some cases, the user wants to change the style of a photograph, as proposed by Bae et al. [21], but she/he does not have in mind a target photograph or it may be difficult to find a suitable one for her/his idea of style. However, the user can express a style idea using keywords to describe it.

An example of systems conveying this idea is AutoStyle [165]. In this system, the user feeds, as input into the system, a photograph to be stylized, I_s, and a target keyword, KW_t. This keyword is passed to Bing Image Search to retrieve a collection of 500 images, which are grouped into clusters

(a)   (b)

Figure 4.26. An example of Rabin et al.'s method [226, 227]: (a) The result of the style transfer method in Figure 4.24 (c). (b) The regularization of the result in (a) after a few iterations. While the fine details are not exaggerated, the style is still successfully transferred.

of images with similar contrast distribution and color spatial layouts. k-means clustering is employed, and for each image the k input is its gist descriptor [204] and low resolution (8×8) of its CIE $L^*a^*b^*$'s a and b channels. At this point, the user has to choose one of these clusters by selecting the cluster representative image. Then, images in the selected cluster, C, are converted from RGB to CIE $L^*a^*b^*$, and the L channel for each image is decomposed into a four-level oversampled Laplacian pyramid; L^1, L^2, L^1, and L^4 such that $L = L^1 + L^2 + L^3 + L^4$. Subsequently, a vignette layer, V, is extracted with a similar parameterization as in Goldman [97] taking into account all L_1 layers of the images in C. V captures the radially symmetric layout of lightness in these images, and it is transferred onto I_s by multiplication, obtaining I_v. At this point, the statistics of the C 3D color histogram are matched to ones of I_v by defining a color transform of each pixel $\mathbf{p}$ as

$$T_p(\mathbf{p}) = \sum_{\mathbf{x}} \frac{s(\mathbf{p}, \mathbf{x})}{\sum_{\mathbf{x}} s(\mathbf{p}, \mathbf{x})} T_b(\mathbf{x})) \qquad s(\mathbf{p}, \mathbf{x}) = e^{\frac{-(\mathbf{x} - \mathbf{p})^2}{2\sigma^2}}, \qquad (4.75)$$

where $\mathbf{x}$ is a color representing the mean pixel value in a histogram bin, $\sigma = 0.04$, and $T_b(\mathbf{x})$ is defined as:

$$T_b(\mathbf{x}) = \mathbf{K}x + \Delta, \qquad (4.76)$$

where $\mathbf{K}$ and Δ are computed through a minimization process that avoids value stretching. After the color transfer, local contrast transfer is applied to the resulting image, I_c. This transfer is performed by 3D histogram transferring in a similar way to color transfer. In this case, the first three

(a) I_s (b) Sepia

(c) Desert   (d) New York

Figure 4.27. An example of AutoStyle [165] results for an input image, I_s, with different keywords, KW_t. These results were generated with the help of Yiming Liu.

levels of the Laplacian pyramid of I_s, L_s^1, L_s^2, and L_s^3, are modeled with a 3D histogram over values in the Laplacian pyramid. Figure 4.27 shows AutoStyle results applied to the same image with different KW_t. AutoStyle was validated by running a user study. The study results suggest that the system can robustly transfer a variety of styles, but it has a problem consistently producing style transfer results with agreement amongst users.

Style Transfer for Headshot Portraits. As discussed in Bae et al.'s work [21], style transfer for portraits is challenging, because detail enhancement can increase skin defects. Therefore, this issue has to be carefully managed. To address it, Shih et al. [249] proposed a system based on a multi-scale local transfer in the CIE $L^*a^*b^*$ color space. The first step of the method is to warp the target portrait, I_t, according to the source portrait, I_s. This is achieved using different techniques such as landmark extraction [241] and affine transformation [132] for the rough alignment, and SIFTFlow [163] to refine it. The chain of transformations for warping I_t onto I_s is defined as the operator W. Once I_t is aligned onto I_s, both images are decomposed

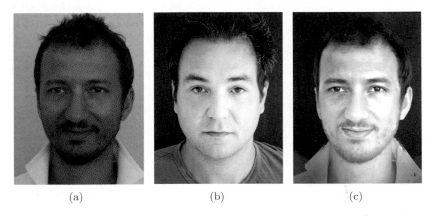

(a)   (b)   (c)

Figure 4.28. An example of portrait style transfer [249]: (a) The source image. (b) The target image. (c) The transfer result. YaChang Shih helped in generating the transfer result.

into six-level Laplacian pyramids [41] without downsampling. Then, for each l-th scale in the pyramids, the local energy, S_l, is estimated as

$$S^l[I] = (\mathbf{L}^l[I])^2 * G(2^{l+1}), \tag{4.77}$$

where $\mathbf{L}$ is the Laplacian operator, I is an input image, and G a Gaussian filter. At this point, local statistics are transferred from I_t to I_s as

$$\mathbf{L}^l[O_s] = \mathbf{L}^l[I_s] \times \text{Gain}^l, \tag{4.78}$$

where O_s is the output image, and Gain^l is the local transfer function which is defined as

$$\text{Gain}^l = \max\left(\min\left(\sqrt{\frac{S^l[W(I_t)]}{S^l[I_s] + \epsilon}}, \theta_h\right), \theta_l\right) * G(\beta 2^l), \tag{4.79}$$

where $\epsilon = 10^{-4}$ is a small value to avoid singularities, $\theta_l = 0.9$ and $\theta_h = 2.8$ are clamping factor, for avoiding artifacts when having large differences, and $\beta = 3$ (found during experiments). Note that the transfer is applied only to a region of interest (i.e., the face) through a mask, previously computed using GrabCut [235]. The negative of this mask is then used to copy the background from I_t to O.

A final important feature to transfer is highlights in the eyes. These are first detected using thresholding (the brightest pixels in an eye), in both $W(I_t)$ and O_s. Secondly, highlights in O are erased and filled using inpainting. Finally, highlights in $W(I_t)$ are composed onto each eye in O_s.

This method was visually compared against Bae et al. [21] and Sunkavalli et al. [258]. The visual comparison clearly shows Shih et al.'s method produces convincing results without overenhancements; see Figure 4.28. The main drawback of the technique is that both persons in I_s and I_t need to share similar facial attributes such as facial hairs, skin color, age, and hair style.

4.4   Summary

In the last 15 years, several approaches have been proposed to transfer color and style from a target image onto a source image. These can be achieved in many different ways such as matching global statistics, local methods, user-based methods, etc. Despite the large number of techniques, this problem is far from being solved; some methods work well for outdoor and indoor scenes but have poor results with human faces, and viceversa. Moreover, few researchers have touched some issues such as HDR content transfer [117, 223, 224]. More importantly, a proper and rigorous methodology for the evaluation of color and style transfer methods is still missing. Typically, evaluation is performed showing visual comparisons, or computing SSIM [283] between the original and transferred luminance channel, or computing the Bhattacharya coefficient [32], or a user study for UI-based methods. However, can a proper methodology be found to assess color and style transfer algorithms? Another interesting fact of color and style transfer is the side research that it has sparked. For example, researchers have recently started to look into how to enhance (retouching) or restore large photograph collections. To solve these problems, many algorithms and tools presented in this chapter have been employed. These are typically coupled with learning [42] or data-driven methodologies [131, 278].

5

Spatial Retargeting

Image content is nowadays visualized on a variety of displays, from the small screen of a cellphone to a large home TV. Many websites display images whose size is dynamically adapted to the browser window size and the surrounding text. All these display options require different formats and resolutions, and the images need to be scaled to fit the available display space. If the aspect ratio of a display is the same as the image, uniform rescaling can be applied, but otherwise, an image needs to be *retargeted* to the display aspect ratio. The two most common strategies are cropping, which uniformly scales the image to fit the entire screen and discards the parts that do not fit, and letterboxing, which adds black bands around the image. To prevent losing parts of the image and at the same time avoid wasting screen space, content-aware retargeting techniques were developed. These methods selectively deform the input image into the target dimensions according to a saliency map, preserving the shape of important image components while distorting unimportant background content (see Figure 5.1). A few general methodologies for retargeting were proposed in recent years, such as discrete carving, continuous warping, and hybrid

Image from [101].

Figure 5.1. Spatial image retargeting enables changing the size and aspect ratio of an image while preserving its visually important parts. For this purpose, a saliency map is computed for the original image (left) and image content is protected from extreme deformation when producing the resized result (right).

Image from [101].

Figure 5.2. Application of image retargeting to photo gallery presentation. A standard way of displaying an image gallery is shown on the top, where much screen space is wasted. On the bottom, all image thumbnails have been retargeted to a uniform size, thereby utilizing the screen space much better and preserving important image content at the same time.

approaches [246] and they are even available in modern commercial image editing software [5].

In this chapter, we introduce the basic problem of spatial image retargeting and give examples of several representative methods to solve it. We present the common pipeline for resizing, used in both discrete and continuous methods: first, an importance or saliency map is computed for the image, and then a retargeting operator is applied while using this map.

We also discuss interactive methods that support fine-tuning of the saliency map in real time to further improve the quality of the retargeting.

To give the reader an idea about spatial retargeting approaches in a nutshell, the basic seam-carving approach on images employs dynamic programming to find the least noticeable seam to be removed. Dynamic programming can be replaced by a graph-cut algorithm, which scales better for larger images. The image warping approach views the image as a continuous domain that can be reshaped by applying a deformation function. Each point in the domain is assigned a local transformation; points in important regions should undergo a shape-preserving transformation, whereas less salient areas can deform more freely. The overall warping function can be computed on a discrete grid that is decoupled from the image resolution: it is thus possible to trade accuracy for speed, developing interactive techniques even for very large images. While seam carving removes image content, warping redistributes its density.

5.1   Problem Statement

The retargeting problem can be stated as follows: Given an image $\mathbf{I}$ of size $n \times m$, and a new size $n^* \times m^*$, we would like to produce a new image $\mathbf{I}^*$ of size $n^* \times m^*$ that will be a good representative of the original image $\mathbf{I}$. However, to date, there is no clear definition or measure as to the quality of $\mathbf{I}^*$ being a good representative of $\mathbf{I}$. In general, there are three main objectives for retargeting:

1. The important *content* of $\mathbf{I}$ should be preserved in $\mathbf{I}^*$.
2. The important *structure* of $\mathbf{I}$ should be preserved in $\mathbf{I}^*$.
3. $\mathbf{I}^*$ should be free of visual *artifacts*.

The definition of *important* is subjective and dependent on the final application. The importance is encoded in an importance function, usually called a *saliency map*, which assigns a scalar value between 0 and 1 to each pixel of an image. Depending on the application, the saliency map can be computed in different ways: it might use low-level visual cues such as image edges, high-level cues such as people's faces, or a combination of these. The advantage of using an importance map is that the retargeting operators become independent from the application-specific knowledge: a retargeting operator takes as input an image, its saliency map, and a set of sizing constraints and produces the final retargeted image.

In the following, we first focus on the retargeting operators and assume a saliency map given for each image, and then discuss automatic and interactive methods to define the saliency map.

Figure 5.3. Cropping an image may remove important image content (middle). Homogeneous scaling alters the size of the objects in the image and may introduce distortion to their proportions (right).

5.2 Basic Retargeting Operators

We first present two simple retargeting operators that are available in many commercial softwares, including the standard image viewers in Microsoft Windows, MacOSX, iOS, and Android. These operators allow us to introduce the notation and then move on to the more complex content-aware operators.

5.2.1 Cropping

The cropping operator is the simplest resizing operator: it simply extracts a window of the desired size $n^* \times m^*$ from the original image. Obviously, this can be done only if $m^* \leq m$ and $n^* \leq n$. Many times, cropping is done manually, i.e., the user chooses a window of the desired size inside the original image. However, given an importance map, one can also search automatically for the window of size $n^* \times m^*$ that contains the most important parts [79, 240]. This does not always produce the expected result; see, for example, Figure 5.3.

A trivial extension of this operator pads the frame with empty pixels, enabling us to enlarge the image size. This is sometimes called *letterboxing* and is ubiquitously used to adapt old TV footage to new screens and vice versa. In terms of the three objectives for retargeting, cropping mostly preserves the structure of the image and does not produce artifacts apart from cutting the image at the fringes. The main downside of using cropping is that content is always lost from the image and the original image composition can be damaged in the process.

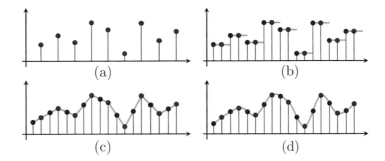

Figure 5.4. Reconstruction and super-sampling of a 1D signal (a) using piecewise-constant (b), linear (c) and cubic (d) interpolation, respectively.

5.2.2 Scaling

Scaling is defined by a homogeneous map between the pixels of the original image $\mathbf{I}$ and the pixels of the target image $\mathbf{I}^*$. By *homogeneous* we mean that all points in the image undergo the same local transformation. If the mapping function is the same for the horizontal and vertical directions, we call the scaling *uniform*. The most straightforward way of implementing this operator is to use forward mapping to transform the pixels of the original image to their new positions, as follows:

$$\mathbf{I}(x,y) \rightarrow \mathbf{I}^*(\lfloor x \cdot \frac{m^*}{m} \rfloor, \lfloor y \cdot \frac{n^*}{n} . \rfloor)$$

However, forward mapping may lead to a many-to-one mapping, as well as to empty pixels in the target image. The trivial solution to this is to use backward mapping:

$$\mathbf{I}^*(X,Y) = \mathbf{I}(\lfloor X \cdot \frac{m}{m^*} \rfloor, \lfloor Y \cdot \frac{n}{n^*} \rfloor).$$

However, this simple approach may lead to either losing some of the original pixels or to their duplication. A better solution is to use a filter: first, a model of the original signal is defined and reconstructed using a given interpolation function of the pixels of $\mathbf{I}$, the original image, e.g., using the standard bi-linear, bi-quadratic or bi-cubic polynomial interpolation functions. The filter is then used to sample the color at the desired target resolution and size to create the target image $\mathbf{I}^*$. See an example for up-sampling a 1D signal in Figure 5.4.

Still, even homogeneous bi-cubic scaling can create artifacts, such as blockiness and aliasing, and important objects could be scaled beyond the point of recognition. It might seem tempting to use this approach as an image retargeting operator to change the aspect ratio of an image by using

as before, we pick the pixel with the lowest summed saliency and we then backtrace up in the rows until we reach the top. Once the entire seam is found, it can be removed to reduce the size of the image. See Figure 5.5 for an illustration.

Repeating this procedure multiple times allows one to arbitrarily reduce the width of an image. The extension of this algorithm to remove rows is straightforward and it only requires us to transpose the image (and its saliency map) before performing the seam carving operation.

Order optimization. If the target size is smaller in both height and width with respect to the original image, the seam carving algorithm has to be applied to remove both horizontal and vertical seams. Clearly, the order in which horizontal and vertical seams are removed affects the final result, since the removal of a vertical seam directly impacts the cost of all the subsequent horizontal seams and vice versa. Not surprisingly, the previous dynamic programming formulation can be extended to this more general case by encoding the choice of removing a vertical or horizontal seam at the i-th iteration in a Boolean variable $\delta_i = \{0, 1\}$. The optimal sequence can then be found by minimizing the following cost:

$$\min_{\mathbf{s}^{\mathsf{x}}, \mathbf{s}^{\mathsf{y}}, \delta} \sum_{i=1}^{k} E\left(\delta_i \, \mathbf{s}_i^{\mathsf{x}} + (1 - \delta_i) \, \mathbf{s}_i^{\mathsf{y}}\right) \quad \text{s.t.} \quad (5.3)$$

$$k = r + c, \quad r = m - m^*, \quad c = n - n^*,$$

$$\delta_i \in \{0, 1\}, \quad \sum_{i=1}^{k} \delta_i = r, \quad \sum_{i=1}^{k} (1 - \delta_i) = c.$$

As before, we accumulate the minimal cost in a table T that assigns to every desired target image $n' \times m'$ the cost of the optimal sequence of seam removals. The entry $T(r, c)$ stores the minimal cost to obtain an image of size $(n - r) \times (m - x)$. The table T is initialized with $T(0, 0) = 0$ and then dynamically filled with the recursion

$$T(r, c) \;=\; \min\left\{ \begin{matrix} T(r - 1, c) + E\left(\mathbf{s}^{\mathsf{x}}(\mathbf{I}_{(n-r-1) \times (m-c)})\right), \\ T(r, c - 1) + E\left(\mathbf{s}^{\mathsf{y}}(\mathbf{I}_{(n-r) \times (m-c-1)})\right) \end{matrix} \right\}, \quad (5.4)$$

where $\mathbf{I}_{(n-r) \times (m-c)}$ denotes an image of size $(n - r) \times (m - c)$, and $E(\mathbf{s}^{\mathsf{x}}(\mathbf{I}))$ and $E(\mathbf{s}^{\mathsf{y}}(\mathbf{I}))$ are the costs of the respective seam removal operations. The intuition behind this recursion is similar to the previous process: for every possible operation, we try it and we store the result with the minimal energy. At the end of the recursion, the entire T is filled with values, and we can find the value with the smallest cost that corresponds to our target size, and then backtrack to find the optimal sequence of seam removal operations.

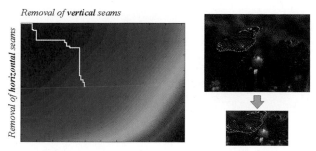

Figure 5.6. Optimal-order retargeting. The original image is shown on the top right, and the transport map T to obtain the desired resized image (bottom right) is shown on the left. Given a target size, the method follows the optimal path from the target size to the origin (upper left corner) and removes either a vertical or a horizontal seam in each step to obtain the retargeted image. This path is indicated in white on T.

Seam duplication. The seam carving algorithm can be inverted to increase the size of an image instead of reducing it. The algorithm is identical to the previous case, but instead of removing a seam, we insert a new one in the same position, such that the new pixels have the average colors of their neighbors. Since the optimal seam is the one with the lowest saliency, it will be placed in an unimportant part of the image. Duplicating this seam thus extends the background of the image without introducing noticeable artifacts. However, care has to be taken when this process is repeated: naively applying this algorithm multiple times would always pick the same seam and result in entire regions of the image with uniform colors. To avoid this, it is sufficient to select the first k seams with minimal energy (k is the number of seams we wish to introduce), and duplicate all of them at the same time.

Enlarging an image with this algorithm can be used as an alternative way of changing the aspect ratio of an image. Instead of changing the aspect ratio by removing rows or columns, it is possible to change it by adding them. This has the advantage of preserving the entire content in the image and avoiding removal of any salient pixel, but it increases the image size.

Variants. A different cost formulation called *forward* seam carving has been introduced in [237] to tackle a subtle but important limitation of the formulation presented in the previous paragraphs: Every time a seam is removed, the saliency map should be recomputed to take into account the saliency of the pixels adjacent to the ones removed. Not taking this change into account leads to suboptimal retargeting results, and the naive approach

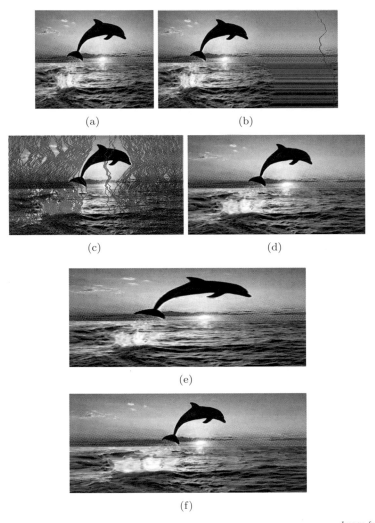

Image from [17].

Figure 5.7. Seam insertion. Naively inserting the optimal seam by duplication leads to the repeated insertion of the same seam (b). Seam insertion in order of their (hypothetical) removal (c) leads to a more visually pleasing result (d). Two steps of seam insertions of 50% in (f) achieves better results than homogeneous scaling (e).

of recomputing the saliency after every seam removal is computationally expensive. Rubinstein and colleagues [237] propose to directly incorporate the saliency changes in the optimization, obtaining an efficient algorithm that accounts for the saliency changes. They also replace the dynamic programming algorithm we discussed here with a more complex graph-cut formulation to further reduce the computational costs and directly support the retargeting of video sequences.

5.4 Continuous Retargeting Operators

Continuous retargeting operators view the image as a continuous function and perform a continuous geometric deformation to fit it into the new desired shape. The image content is virtually transformed by a continuous deformation, or warp, and then discretized by resampling the image signal inside the target domain. The resampling, similar to the simple scaling operator, is usually performed using bi-cubic interpolation. In fact, the scaling operator is the simplest continuous retargeting operator, which is restricted to using the same deformation (a uniform scaling) for all parts of the image. Modern continuous spatial retargeting approaches generalize this idea to arbitrary deformations, under the sole assumption that the boundary of the image must be mapped to the boundary of the target domain. To improve the quality of the retargeted result, the salient parts of the image should undergo the least distortion, while concentrating the distortion in the low-importance regions. A continuous retargeting operator tries to compute such a warp, and this is usually achieved using a variational method: The deformation is parameterized using a finite set of variables (usually, the vertices of a regular grid over the image) and the quality criteria of the map are encoded in an energy function. The energy is minimized with a numerical optimization method, which computes the optimal deformation, which is then used to produce the retargeted image.

In this section we describe the continuous theory and a discretization using a mesh, and then we list the most common energies in the state of the art. For clarity, we use the same notation as in [148, 282].

Setup and notation. As everywhere throughout this chapter, assume that a saliency map $S : \mathbf{I} \to [0, 1]$ is given (see Section 5.5 for more details). We are looking for a warping function $F : \mathbb{R}^2 \to \mathbb{R}^2$ that transforms the input image $\mathbf{I}$ of dimensions $n \times m$ into the output image $\mathbf{I}^*$ with the desired dimensions $n^* \times m^*$. We denote the horizontal component of F by F_x and the vertical one by F_y, such that $F(x, y) = (F_x(x, y), F_y(x, y))$. The warp must transform the boundary rectangle of the input image into the new dimensions, so assuming that we attach the coordinate $(0, 0)$ to the bottom left corner of the image and the coordinate (m, n) to the top right corner, F should satisfy

$$F_x(0, \cdot) = 0, \quad F_x(m, \cdot) = m^*, \quad F_y(\cdot, 0) = 0, \quad F_y(\cdot, n) = n^*.$$

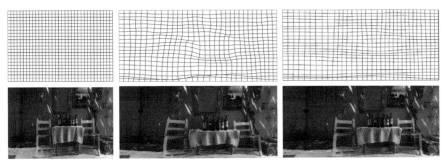

Images from [282].

Figure 5.8. Demonstration of the effect of the line-bending energy term in the scale-and-stretch warping technique of [282]. From left to right: the original image, the retargeted image without and with the line bending energy. Note how adding the energy term prevents strong grid line bending and produces a less distorted retargeting result.

The above equations are the *boundary constraints*; the optimization will search for the best possible F that exactly satisfies those constraints. The "best" behavior is typically defined in terms of the local behavior of F, i.e., its Jacobian

$$J_F(x,y) = \left(\frac{\partial F}{\partial x} \frac{\partial F}{\partial y} \right) = \begin{pmatrix} \partial F_x/\partial x & \partial F_x/\partial y \\ \partial F_y/\partial x & \partial F_y/\partial y \end{pmatrix}.$$

The Jacobian $J_F(x,y)$ is a linear transformation (scaling, shearing, etc.) that best approximates F in a small neighborhood around the point (x,y). Ideally, the Jacobian would equal the identity matrix everywhere, which would mean that F does not distort the image at all. However, this is of course impossible if the size of the image is to be changed, so a more flexible objective functional needs to be defined. For example, one would like the regions of high importance to be distorted the least while sacrificing other regions. This means we could ask the Jacobian of each point (x,y) to be as close as possible to the identity matrix I in the least squares sense, weighted by the importance $S(x,y)$. This results in the following objective functional:

$$E(F) = \int_{x=0}^{m} \int_{y=0}^{n} S(x,y) \left\| J_F(x,y) - I \right\|^2 dx dy, \qquad (5.5)$$

where $\| \cdot \|$ is the Frobenius matrix norm. Written out more explicitly,

$$\left\| J_F(x,y) - I \right\|^2 = \left(\frac{\partial F_x}{\partial x}(x,y) - 1 \right)^2 + \left(\frac{\partial F_y}{\partial x}(x,y) \right) \right)^2 +$$
$$\left(\frac{\partial F_x}{\partial y}(x,y) \right)^2 + \left(\frac{\partial F_y}{\partial y}(x,y) - 1 \right)^2.$$

There exist many variations on the particular terms used in the objective functional, as we will see below, but they all typically involve the partial derivatives of F. The goal is to find the optimal F:

$$F = \underset{F}{\arg\min}\, E(F) \;\; \text{subject to the boundary constraints.}$$

Discretization. So far we have formulated everything in a continuous manner, but in order to compute the optimal warp F, it is usually necessary to discretize the problem so that standard numerical optimization methods can be applied. This is fairly easy thanks to the regular structure of the image domain. Typically, a grid (quad) mesh is superimposed against the image and the discrete objective functional is defined using the mesh vertices. The mesh can have an arbitrary resolution: it can coincide with the pixel grid (or even be finer), but often for the sake of efficiency, a coarser grid is used. Let us denote the grid mesh by $\mathbf{M} = (\mathbf{V}, \mathbf{E}, \mathbf{Q})$ with vertices $\mathbf{V}$, edges $\mathbf{E}$, and quad faces $\mathbf{Q}$, where $\mathbf{V} = [\mathbf{v}_0^T, \mathbf{v}_1^T, ...\mathbf{v}_{end}^T]$ and $\mathbf{v}_i \in \mathbb{R}^2$ denote the initial vertex positions. The vertices and edges form horizontal and vertical grid lines partitioning the image into quads. The problem is then to find the deformed mesh geometry $\mathbf{V}' = [\mathbf{v}_0'^T, \mathbf{v}_1'^T, ...\mathbf{v}_{end}'^T]$, i.e., $F(\mathbf{v}_i) = \mathbf{v}_i'$ for each i. Once the discrete F is computed for the vertices, the image content within each face of the mesh can be reconstructed by interpolation.

It is generally assumed that all the faces of the mesh are square with fixed unit edge length; therefore the partial derivatives of F can be easily discretized using finite differences:

$$\frac{\partial F}{\partial x}(\mathbf{v}_i) = \mathbf{v}_j' - \mathbf{v}_i',$$

where vertex j is the right-hand horizontal neighbor of vertex i, and similarly for the vertical direction,

$$\frac{\partial F}{\partial y}(\mathbf{v}_i) = \mathbf{v}_k' - \mathbf{v}_i',$$

where vertex k is the top vertical neighbor of vertex i.

The importance map S is also discretized and lumped to the mesh vertices. Since S is usually defined on the discrete digital image domain to begin with, if the grid mesh does not coincide with the pixel grid (e.g., the mesh is coarser than the image resolution), then one can take the average of the pixels' importance values in some fixed-size neighborhood of each mesh vertex $\mathbf{v}_i$ to define $S(\mathbf{v}_i)$.

Using such discretization, and replacing integrals with sums, the example objective functional we saw earlier in Eq. (5.5) becomes

$$E(F) = \sum_{\{i,j\}\in\mathbf{E}} S(\mathbf{v}_i)\|(\mathbf{v}_j' - \mathbf{v}_i') - (\mathbf{v}_j - \mathbf{v}_i)\|^2.$$

original image   [237]   [292]   [282]

Figure 5.9. Comparison of warp-based and discrete retargeting techniques.

Explanation: the objective functional would like to keep all the horizontal derivatives equal to $(1, 0)^T$ and all the vertical derivatives equal to $(0, 1)^T$; of course all the original horizontal neighbor vertices $\mathbf{v}_i, \mathbf{v}_j$ satisfy $\mathbf{v}_j - \mathbf{v}_i = (1, 0)^T$, and the vertical neighbors have $\mathbf{v}_j - \mathbf{v}_i = (0, 1)^T$, so it is convenient to compactly write the energy as above, without distinguishing vertical and horizontal cases, simply as $(\mathbf{v}'_j - \mathbf{v}'_i) - (\mathbf{v}_j - \mathbf{v}_i)$.

In the functional above, the unknowns are the warped vertex locations $\mathbf{v}'_i$; note that the functional in this case is quadratic in the unknowns, and thus we can find the warp by solving a sparse linear system of equations (for more details, see [90]). This can be done robustly and quite efficiently with modern numerical solvers; all warp-based retargeting methods try to avoid more involved, nonlinear functionals when possible to keep the computational costs low and the implementation simple.

5.4.1 Warp-Based Image Retargeting

Given the basics described above, let us now look at several representative examples of warp-based spatial retargeting methods. Most techniques fall into the general optimization framework described above and differ by the particular objective functionals they formulate.

Gal et al. [90] were among the first to propose non-homogeneous warps for images. They used a simple binary, hand-drawn importance map S, and the objective functional for the warping function was accordingly divided into two parts: the Jacobian of "important" points should be as close as possible to scaled identity, and the Jacobian of "unimportant" points is allowed to be the standard homogeneous (nonuniform) scaling A that fits

the new dimensions of the image. The objective functional can be written
as

$$E_{[90]}(F) = \iint S(x,y)\|J_F(x,y) - sI\|^2 + (1 - S(x,y))\|J_F(x,y) - A\|^2.$$

Gal et al. [90] noted that uniform scaling of important regions may be
beneficial, as it preserves the shape, and thus the image content, yet allows
more flexibility in the warp. They simply defined the uniform scaling factor
s to be the minimum between the horizontal and vertical scaling induced by
the new dimensions of the image (so for instance, if the image is stretched
or shrunk just along one dimension, the uniform scaling factor would be 1).

The idea of taking advantage of local scaling was taken further by
Wang et al. [282]. They suggested using varying uniform scaling factors for
each point in the image and finding those by optimization (hence naming
the technique "optimized scale-and-stretch"). In discrete form, the initial
objective functional then has the form

$$E_{[282]}(F) = \sum_i S(\mathbf{v}_i) \sum_{j \text{ s.t. } \{i,j\} \in \mathbf{E}} \|(\mathbf{v}'_j - \mathbf{v}'_i) - s_i(\mathbf{v}_j - \mathbf{v}_i)\|^2.$$

Both the optimal new mesh vertices $\mathbf{v}'_i$ and the uniform scaling factors
s_i are unknowns; they are computed by alternating optimization. Fixing
s_i allows us to find the optimal warped vertex positions $\mathbf{v}'_i$ by solving a
sparse linear system; given the current $\mathbf{v}'_i$, the scaling factors s_i can be then
updated, and the iterations repeat until convergence (see the paper [282]
for details).

In addition to this energy functional, Wang et al. [282] also suggest
adding a grid line bending energy: only having the scaling functional above
may cause the grid lines to arbitrary bend, so to limit the "wild" behavior,
it is necessary to encourage the grid lines to keep their original orientation
while allowing the length to change (see Figure 5.8). The bending energy is

$$\sum_{\{i,j\} \in \mathbf{E}} \|(\mathbf{v}'_j - \mathbf{v}'_i) - l_{ij}(\mathbf{v}_j - \mathbf{v}_i)\|^2. \tag{5.6}$$

Here, again, the length factors l_{ij} are unknowns and are iteratively found
during the alternating optimization. In fact, the lengths l_{ij} can be written
as nonlinear expressions in $\mathbf{v}'_i$:

$$l_{ij} = \|\mathbf{v}'_i - \mathbf{v}'_j\| / \|\mathbf{v}_i - \mathbf{v}_j\|.$$

By employing the alternating optimization strategy, i.e., fixing the length
factors, solving for the vertex positions and then updating the length factors
by simply applying the formula above with the current vertex positions $\mathbf{v}'_i$,

it is possible to avoid a complex nonlinear optimization and robustly reach
a reasonable solution within just a few iterations.

In the work of Wolf et al. [292], scaling is performed along one dimension
only (i.e., the image points are encouraged to move in the direction of
the stretch but not orthogonal to it); this is achieved by formulating an
objective functional that asks, without loss of generality, the horizontal
derivatives of the warp to equal 1 while smoothing the vertical derivatives.
Their objective functional thus has the form

$$E_{[292]}(F) = \iint S(x,y) \left(\frac{\partial F_x}{\partial x}(x,y) - 1 \right)^2 + w \left(\frac{\partial F_y}{\partial x}(x,y) \right)^2 dxdy.$$

Wang et al. [282] note, however, that this formulation may often lead to
self-intersections and is sometimes sub-optimal in terms of effective use of
the image domain, because it does not allow features to scale down in both
directions.

In a more recent work, Krähenbühl and colleagues [148] note that
allowing different parts of the image to scale differently can often lead to
significant changes of proportions, so they opt to replace the local scaling
factors s_i by one global scaling factor s, which is also found by alternating
optimization. They also propose two additional energy terms that improve
the appearance of edges:

$$\iint S_e(x,y) \left(\left(\frac{\partial F_x}{\partial y} \right)^2 + \left(\frac{\partial F_y}{\partial x} \right)^2 \right) dxdy,$$

$$\iint S_e(x,y) \left(\left(\frac{\partial F_x}{\partial x} - 1 \right)^2 + \left(\frac{\partial F_y}{\partial y} - 1 \right)^2 \right) dxdy.$$

Here, $S_e(x,y)$ is an edge saliency map (computed by running, e.g., a Sobel
operator); the first term prevents bending of features (similar to [292]) and
the second term prevents edge blurring by enforcing similar image gradients
on feature edges.

In addition to the automatic warping, the work by Krähenbühl et al. [148]
explores various interactive control mechanisms for the retargeting process,
allowing users to interactively mark objects in the image (or video) and
influencing their location, as well as manual markup of lines and edges, which
are then constrained to remain straight. Note that automatic edge bending
energies (like in Eq. (5.6) or above) can only prevent local bending of features,
but do not prevent global distortion of feature lines. A manual line constraint
is created by drawing a line represented as $l : sin(\alpha)x + cos(\alpha)y + b = 0$.
Each image point is then associated with a value $c(x,y)$ that indicates the
coverage percentage by the line; the line preservation energy term can be

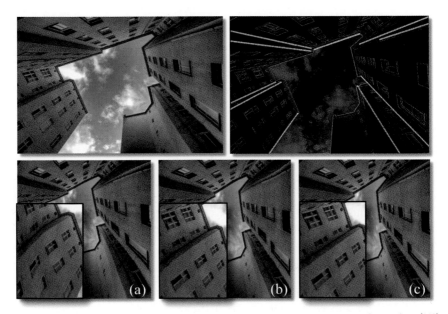

Images from [148].

Figure 5.10. Demonstration of feature edge retargeting. The top row shows the original frame (left) and the edge saliency map S_e (right) used in [148]. Manually added global line constraints are marked in white. The bottom row shows different retargeting results: (a) Wang et al. [282], (b) Krähenbühl et al. [148] with automatic line bending energy terms only, (c) Krähenbühl et al. [148] using the additional manual line constraints.

simply formulated as

$$\iint c(x,y)\left(\sin(\alpha)F_x(x,y)+\cos(\alpha)F_y(x,y)+b\right)^2 dxdy.$$

Note that initially the line parameters α and b are set as in the original image, but they are updated after each iteration in an alternative optimization manner, just like the scaling parameter. See Figure 5.10, which shows the effect of the automatic and manual line constraints.

Some warp-based image retargeting results are shown in Figure 5.18. It is interesting to compare the results of the different warping techniques, and the discrete methods as well: generally, warp-based retargeting does not suffer from discontinuity artifacts typical for seam carving, but the trade-off may be more significant distortion of image objects, smearing artifacts and, when spatially varying scaling is involved, change in relative proportions.

Recently, the finite element method with adaptive domain triangulation has been proposed as a retargeting operator by Kaufmann et al. [139], leading to a generic framework that can produce results of quality that is comparable or even higher than the methods discussed above. In particular, this work showed that the continuous spatial retargeting operators do not have to be discretized over uniform quad grids and may benefit from content-adapted grids or triangle meshes with varying density.

5.4.2   Axis-Aligned Image Retargeting

All the resizing energies we saw above require a few seconds to minimize on a standard workstation when applied to HD images, and preclude the usage of the respective resizing methods on mobile and other devices with limited memory and computational power. Observing the behavior of the state-of-the-art warping methods we discussed, one can notice that they hardly introduce any local rotations in the image deformation. This indeed makes sense, since if the resizing operator contains local, varying rotations, they manifest themselves as "swirls," which are highly noticeable distortion artifacts. Absence of local rotations leads the retargeting deformation to be *axis-aligned*, i.e., the isoparametric lines remain straight and parallel after deformation, only changing the spacing between themselves (see Figure 5.11). A key observation to drastically reduce the complexity of a continuous retargeting operator is therefore to restrict its deformation space to the space of *the space of axis-aligned deformations*.

This observation has important consequences. First, the deformation space can be parameterized in 1D, since an axis-aligned deformation is determined by the intervals between the vertical and horizontal isoparametric lines. Previous retargeting methods parameterize the deformations in 2D, leading to optimization problems in the order of $m \cdot n$ unknowns, whereas a 1D parameterization necessitates only $O(m + n)$ unknowns. Further, preventing foldovers and controlling the stretching of axis-aligned deformations is simple and robust, since it merely poses linear inequality constraints on the isoparametric line spacing. Finally, axis-aligned deformations open up the possibility of employing standard elastic deformation energies, which do not inherently penalize local rotations and were therefore ineffective for previous, 2D-parameterized warping methods.

As before, we overlay a uniform grid over the image with m columns and n rows and assume unit width and height for each column and row. The task is to compute a deformed grid for the resized image, with the desired total width m^* and height n^*. In the continuous setting, an *axis-aligned deformation* can be fully described by the vertical and horizontal deformation derivatives along the boundary. In our discrete setting, we assume an axis-aligned deformation to be piecewise-linear (linear on each grid cell), such that it is fully determined by the widths of the deformed grid columns and the heights of the deformed grid rows.

Let $\mathbf{s}^{\text{rows}} = (s_1^{\text{rows}}, s_2^{\text{rows}}, \ldots, s_n^{\text{rows}})$ denote the unknown heights of the

[138]   [50]

[282]   [208]

Figure 5.11. Excluding local rotations leads the retargeting deformation to be *axis-aligned*. The *Dogs* dataset is retargeted to 50% width using recent techniques and our method. Observe that although previous approaches do not explicitly enforce it, their deformations are nearly axis-aligned.

rows and $\mathbf{s}^{\mathrm{cols}} = (s_1^{\mathrm{cols}}, \ldots, s_m^{\mathrm{cols}})$ the unknown widths of the columns. The axis-aligned deformation is therefore represented by the vector of unknowns $\mathbf{s} = (\mathbf{s}^{\mathrm{rows}}, \mathbf{s}^{\mathrm{cols}})^T \in \mathbb{R}^{m+n}$. The general form of the optimization that computes the retargeted image grid is

$$\text{minimize} \quad \mathbf{s}^T Q \mathbf{s} + \mathbf{s}^T \mathbf{b} \tag{5.7}$$

$$\text{subject to} \quad s_i^{\mathrm{rows}} \geq L^h, \quad i = 1, \ldots, n, \tag{5.8}$$

$$s_j^{\mathrm{cols}} \geq L^w, \quad j = 1, \ldots, m, \tag{5.9}$$

$$s_1^{\mathrm{rows}} + \ldots + s_n^{\mathrm{rows}} = n^*, \tag{5.10}$$

$$s_1^{\mathrm{cols}} + \ldots + s_m^{\mathrm{cols}} = m^*. \tag{5.11}$$

$Q \in \mathbb{R}^{(m+n) \times (m+n)}$ and $\mathbf{b} \in \mathbb{R}^{m+n}$ are determined based on the particular

original   ASAP   ARAP

Images from [208].

Figure 5.12. The *Fatem* image resized with different energies.

deformation energy functional, and $L^h, L^w > 0$ are the minimum sizes allowed for rows and columns of the deformed grid. The inequalities (5.8) and (5.9) *guarantee* that our deformation is free of foldovers, since every cell on the grid cannot be smaller than the specified dimensions L^h-by-L^w and cannot invert (inside each cell the deformation is linear and therefore foldover-free). Equations (5.10) and (5.11) fix the total dimensions of the deformed grid to the desired target size.

For the above quadratic program (QP) to be feasible, we just need $L^h \leq n^*/n$ and $L^w \leq m^*/m$; simple homogeneous scaling then provides a feasible solution. The feasible domain is bounded, since $\forall i, 0 \leq s_i \leq \max\{m^*, n^*\}$, such that the objective function in (5.7) is finite in the feasible region. The energy should be defined in such a way that Q is positive (semi)definite; our problem is then convex and can be solved with standard QP solvers.

5.4.3 Energy Functionals for Axis-Aligned Retargeting

To integrate the saliency map $S(\cdot)$ in the above formulation, its values are averaged inside every cell of the grid on the original image, obtaining the saliency matrix $S \in \mathbb{R}^{n \times m}$.

While any of the energy functionals $E(F)$ described previously can be minimized in the space of axis-aligned deformations, we describe two additional ones below, the *as-similar-as-possible* (ASAP) energy [306] and the *as-rigid-as-possible* (ARAP) energy [253]. These are often used in graphics literature to design freeform deformations for shape modeling and animation purposes, but as mentioned previously, they do not penalize local rotations at all and are hence problematic for image resizing. The axis-aligned deformation space gives these energies a renaissance and makes them highly useful. Figure 5.12 shows an example result with these energies.

ASAP Energy. In the space of axis-aligned deformations, a similarity transformation is a combination of uniform scaling and translation, since rotations are eliminated. The ASAP energy thus minimizes nonuniform scaling:

$$E_{\text{ASAP}} = \sum_{i=1}^{n} \sum_{j=1}^{m} \left(S_{i,j} \left(s_i^{\text{rows}} - s_j^{\text{cols}} \right) \right)^2. \qquad (5.12)$$

To minimize this energy, we define the following matrix $K \in \mathbb{R}^{(mn) \times (m+n)}$:

$$K_{k,l} = \begin{cases} S_{r(k),c(k)} & \text{if } l = r(k), \\ -S_{r(k),c(k)} & \text{if } l = n + c(k), \\ 0 & \text{otherwise,} \end{cases} \qquad (5.13)$$

where $r(k) = \lceil k/m \rceil$ and $c(k) = ((k-1) \bmod m) + 1$. From this equation, $K\mathbf{s}$ gives us the vector with energy terms per row, and $E_{\text{ASAP}} = \mathbf{s}^T K^T K \mathbf{s}$. Using the generic notation of Eq. (5.7), $Q = K^T K$ and $\mathbf{b} = 0$. Clearly, Q is a positive semidefinite matrix, such that the energy is convex.

ARAP Energy. In the axis-aligned deformation space, a rigid transformation is reduced to a translation, since rotations are not allowed by definition. The ARAP energy thus minimizes uniform and nonuniform

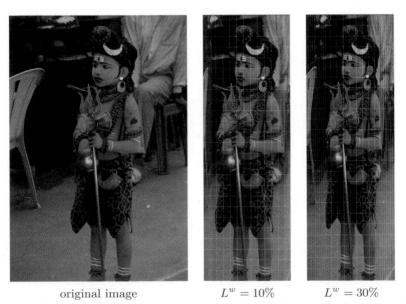

original image $L^w = 10\%$ $L^w = 30\%$

Images from [208].

Figure 5.13. The minimal column width L^w and row height L^h can be prescribed. Middle: each column may not be compressed to more than 10% of the original width. Right: the minimal width is 30%, such that the deformation is less pronounced, since extreme squeezing is disallowed.

original   saliency   grid size 10×10   25×25   50×50   100×100

Images from [208].

Figure 5.14. Grid resolution does not have a dramatic effect on the energy minimization result (here, the ASAP energy was used).

scaling:

$$E_{\text{ARAP}} = \sum_{i=1}^{M} \sum_{j=1}^{N} S_{i,j}^2 \left((s_i^{\text{rows}} - 1)^2 + (s_j^{\text{cols}} - 1)^2 \right).$$

To minimize this energy, we define the following two matrices $R^{\text{top}}, R^{\text{btm}} \in \mathbb{R}^{(mn) \times (m+n)}$:

$$R_{k,l}^{\text{top}} = \begin{cases} S_{r(k),c(k)} & \text{if } l = r(k) \\ 0 & \text{otherwise,} \end{cases} \tag{5.14}$$

$$R_{k,l}^{\text{btm}} = \begin{cases} S_{r(k),c(k)} & \text{if } l = M + c(k) \\ 0 & \text{otherwise,} \end{cases} \tag{5.15}$$

where $r(k) = \lceil k/m \rceil$ and $c(k) = ((k-1) \bmod m) + 1$. We also define the vector $\mathbf{v} \in \mathbb{R}^{mn}$, $v_k = S_{r(k),c(k)}$. We can now rewrite the ARAP energy using matrix notation:

$$E_{\text{ARAP}} = \left(\begin{bmatrix} R^{\text{top}} \\ R^{\text{btm}} \end{bmatrix} \mathbf{s} - \begin{bmatrix} \mathbf{v} \\ \mathbf{v} \end{bmatrix} \right)^T \left(\begin{bmatrix} R^{\text{top}} \\ R^{\text{btm}} \end{bmatrix} \mathbf{s} - \begin{bmatrix} \mathbf{v} \\ \mathbf{v} \end{bmatrix} \right). \tag{5.16}$$

In the generic notation of Eq. (5.7):

$$Q = \begin{bmatrix} R^{\text{top}} \\ R^{\text{btm}} \end{bmatrix}^T \begin{bmatrix} R^{\text{top}} \\ R^{\text{btm}} \end{bmatrix}, \quad \mathbf{b} = -2 \begin{bmatrix} R^{\text{top}} \\ R^{\text{btm}} \end{bmatrix}^T \begin{bmatrix} \mathbf{v} \\ \mathbf{v} \end{bmatrix}. \tag{5.17}$$

Similar to the ASAP energy case, we end up with a quadratic programming problem with a small positive semidefinite matrix, such that the whole resizing problem can be very efficiently solved using standard QP solvers. In fact, Graf and colleagues [101] show that coarse grid resolutions can be employed at no significant detriment to the result quality (see Figures 5.14, 5.15), and the speedup leads to real-time frame rate performance of the resizing operator on mobile devices such as smartphones. This enables, among other things, interactive update of the saliency map S by the user in real time (more on saliency computation below).

original
(detail)   bilinear interpolation   B-spline interpolation

Images from [208].

Figure 5.15. Images of high resolution may benefit from higher-order interpolation when using coarse grids for the retargeting optimization. The input image resolution is 2800×1800 and the retargeting uses a 25×25 grid. Bilinear interpolation leads to some smoothness artifacts (middle), while upsampling to a 100×100 grid using the spline technique described in [208] results in visually smooth interpolation.

Original   AA   QP   MO   SV   SC

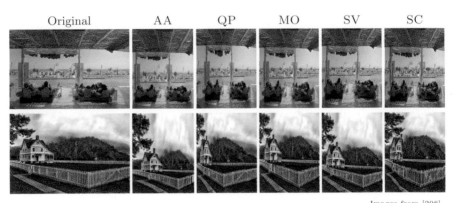

Images from [208].

Figure 5.16. Comparisons between 2D and axis-aligned spatial retargeting methods: AA [208], QP [50], MO [238], SV [148], and SC [237].

5.5   Saliency Measures

All image retargeting operators require a saliency map to decide how to distribute the distortion and which important regions of the image to preserve. We denote the saliency map as a function $S : \mathbf{I} \to [0, 1]$. This map ranks every pixel in an image according to its visual importance. There are different visual cues that affect our perception of visual content, some of them are low-level features such as edges (detected by, e.g., high local intensity gradients), and others are high-level features, for example faces, structures, and symmetries.

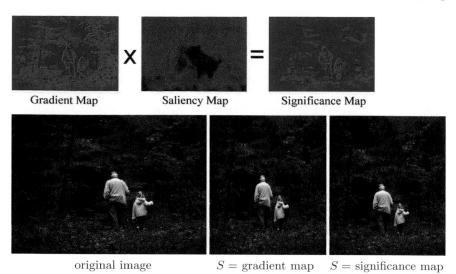

| Gradient Map | Saliency Map | Significance Map |

original image S = gradient map S = significance map

Images from [282].

Figure 5.17. Various importance maps. The top row shows the magnitude of gradients map and Itty's saliency map [127]; these two are combined by multiplication to obtain visual importance in [282]. The bottom row displays retargeting results with the gradient map alone, and using the combined map; adding the multiresolution saliency measure helps filter out spurious gradients in the foliage area and leads to a better result.

5.5.1 Low-Level Features

A common low-level saliency measure, which we denote by S_1, is in fact an edge map, computed simply from the magnitude of intensity gradients and then normalized to $[0, 1]$:

$$S_1(\mathbf{I}) = \left|\frac{\partial}{\partial x}\mathbf{I}\right| + \left|\frac{\partial}{\partial y}\mathbf{I}\right|. \tag{5.18}$$

The rationale behind using edges is to preserve strong contours in the image, as those usually delineate prominent objects. Furthermore, human vision is more sensitive to edges, and this importance map assigns a high saliency value to edges and a low saliency value to smooth areas. Instead of using the L_1 norm to measure the error, some works prefer the L_2 norm:

$$S_2(\mathbf{I}) = \sqrt{\left(\frac{\partial}{\partial x}\mathbf{I}\right)^2 + \left(\frac{\partial}{\partial y}\mathbf{I}\right)^2}. \tag{5.19}$$

Another common measure of local importance is image entropy $S_3(\mathbf{I})$, which can be defined in every image pixel (i, j) as the entropy of a small $(k \times k)$

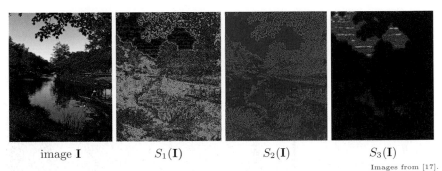

image **I** $S_1(\mathbf{I})$ $S_2(\mathbf{I})$ $S_3(\mathbf{I})$

Images from [17].

Figure 5.18. From left to right: The original image, its saliency map computed with the L_1 and L_2 norms of the gradient magnitude, and with image entropy.

window around that pixel:

$$(S_3(\mathbf{I}))_{i,j} = -\sum_l p_l \log_2(p_l), \qquad (5.20)$$

where p_l are the image intensity histogram values in the $k \times k$ window and l are the histogram bins.

The three local measures are compared in Figure 5.18, where they all tag the uniform sky with a low saliency value, while favoring the vegetation since it is rich in high-frequency details.

Other edge detection mechanisms such as the Canny edge detector could also be used. However, strong edges might sometimes appear also in noisy regions, which are not necessarily salient. A more elaborate image saliency measure was developed, for instance, by Itty et al. [127]. They build a multiresolution pyramid of the image and look for significant intensity and color changes on all levels, combining those into a single, high-resolution map. Other possible low-level measures are the Harris corners measure [109], histograms of gradients (HoG) [59], and entropy, which all measure some properties of a window around each pixel. More complex local descriptors analyze the global contrast of an image [51, 214] or Fiedler vectors [215].

Not surprisingly, it was shown by Wang et al. [282] that combining the gradient map and the saliency map together (by multiplication) has benefits over using just a single measure, since Itty's saliency filters out noisy gradients (see the example in Figure 5.17).

A radically different approach has been proposed in [104, 116], where the spectral properties of images are used to detect "anomalies" that usually correspond to salient objects. The key observation is that, in a large collection of images, the average log spectrum suggests a local linearity; the bumps in the graphs (Figure 5.19) are likely to represent anomalies that are caused by salient objects. This approach has the advantage of not requiring

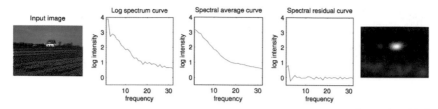

Figure 5.19. The spectral residual can be converted in a saliency map [116].

Figure 5.20. From top to bottom: input image, salient map computed by [127], and salient map computed by [164].

any training data or description of features, but has the disadvantage of producing blurry saliency maps.

A machine learning approach to computing saliency has been recently proposed in [164], where a set of features such as multi-scale contrast, center surround histogram, and color spatial distribution are used to describe salient objects manually annotated in a big image database. A conditional random field is learned to effectively combine these features and used for salient object detection (Figure 5.20). This method produces accurate salience maps with the salient object segmented from the background, but it is sensitive to high-frequency content like edges and noise and it can only be applied in images with one single salient object.

Another category of local methods is based on the concept of local similarity between image patches [96]. For each pixel, a combinatorial search is performed over the image to find similar neighborhoods. If many

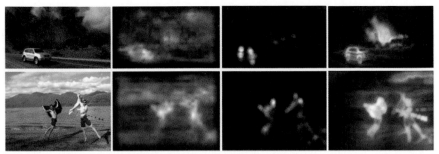

Figure 5.21. From left to right: Input images, saliency computed with [127], the spectral method [116] and the patch-based approach proposed in [96].

positives are found, then the pixel has low importance since it is similar to a large part of the image that is likely to be the background. This search is performed at different scales with patches of varying sizes. An example is shown in Figure 5.21, where [96] is compared with other local methods.

5.5.2 High-Level Features

Some high-level important features can be detected automatically, for example, human faces are often detected using [275]. Depending on the application, the user may choose different high-level features that are important and use them to compute a saliency map. Often, the user may have a specific idea about which content is important, and this may not necessarily correspond to the automatic measures. A possible solution is to measure the eye gaze of real viewers [65] and infer the more important (salient) parts in the image [240]. The gaze information can be used to construct a model that is then used to automatically create a saliency map for new images [134] or as an objective criterion to evaluate the quality of retargeting results [44].

Structural properties of an image that are usually preserved by modern retargeting operators are straight lines [45, 149] and symmetry [294]. Combining these methods with a learning approach that considers both local and global descriptors can further improve the quality of the saliency map [303], as shown in Figure 5.22.

Many recent works also allow some direct degree of user intervention either by designating specific structural constraints that should be preserved [148] or even directly drawing on the importance map to assign importance values to pixels [17, 102, 208].

Preserving all the important features at once may be a difficult, if not

Figure 5.22. From left to right: input images, saliency map computed with [127], the spectral method [116], and the saliency maps obtained with [303]. The last two images were also automatically tagged with the labels "Bike" and "Person," respectively.

infeasible task, as the different constraints might conflict with each other. The proper weighting of all the different measures against each other is a challenge and highly depends on the semantics of the image content itself. Therefore, as mentioned before, requiring some amount of user intervention may be unavoidable at times for acceptable results.

5.6   Summary

In this chapter, we presented the general approaches to spatial image retargeting and summarized some representative methods. The content-aware image resizing methods can be classified into two rough categories: discrete and continuous approaches. While the algorithms used are quite different, clear common grounds and parallels exist: Both types of methods try to achieve the best possible resizing result by optimizing an appropriate energy functional, and they do this by sacrificing unimportant visual content in order to leave room for well-preserved salient visual information. Both types of approaches have their advantages and disadvantages: broadly speaking, discrete methods generalize cropping and thus handle removal of unnecessary content well, which is especially evident for high-frequency, textured image content (such as foliage, sand, water, etc.). Continuous approaches tend to avoid discontinuity artifacts and typically preserve the overall shapes of image objects more coherently. Interestingly, some continuous methods do not heavily penalize extreme shrinking of unimportant image regions, in which case these regions may shrink to nearly vanishing width, effectively resulting in complete content removal, just like in the discrete methods.

Current research in spatial image retargeting is mainly focused on improving the efficiency of retargeting, to enable its usage on mobile devices and on extending these techniques to video content, as this area has extremely high impact in terms of applications and is considerably more

challenging than retargeting a single image. Additionally, perceptual studies of media retargeting with human subjects should play an important role
in further progress. A few recent works performed some experiments of this
kind [44, 148, 236, 238], in particular, Rubinstein and colleagues [236], have
established a benchmark called RETARGETME for comparisons and a user
study protocol. However, there remains a considerable amount of questions
to be studied, such as: What do people like and dislike in transformed
images and videos? Which artifacts affect this judgment, and which factors
contribute to the overall perception of the spatial retargeting results? An
important avenue for future research would be to come up with a perceptually significant and validated visual metric, suitable for the evaluation of
spatial retargeting methods.

6

Quality Assessment (QA)

Principled quality assessment methods are necessary for evaluating the advantages and disadvantages of the content retargeting techniques we have presented so far in this book. In a perfect world with unlimited time and resources, we would assess the quality of a new retargeting method by designing and performing subjective studies utilizing tried and tested experimental protocols. The main advantage of relying on test subjects is, that instead of devising algorithms for assessing quality, we can utilize the subjects' mental faculties and simply explain to them how we would like them to judge quality. For most real-world scenarios, however, subjective experiments are not practical to execute for reasons we discuss later (Section 6.2).

Consequently, objective quality assessment has been an active area of research where the goal is developing algorithms that can predict content quality automatically without human involvement. The accuracy of the subjective quality rating predicted by an automatic quality assessment method can be judged by how well it correlates with the corresponding subjective rating. At the time of this writing, objective quality assessment methods are moderately successful at predicting subjective quality on certain research data sets. However, we still have a long way to go until objective metrics can reliably estimate content quality in the wild.

Usually, relatively high-precision objective quality assessment procedures involve the use of the *reference* content in addition to the *test* content. The reference content does not contain any visual artifacts, and thus, if the test content has perfect quality, it is visually equivalent to the reference content. This way, the quality assessment problem is reduced to determining the loss of fidelity of the test content with respect to the reference content.

In case of *retargeting quality assessment*, the reference content refers to the original source content, whereas the test contents are obtained using the retargeting methods that are being evaluated. While the simultaneous use of the reference and test contents is not necessarily a problem for many quality assessment tasks, it becomes a significant challenge when the test content is retargeted. This challenge stems from the test and reference contents being fundamentally different. For example, in aspect ratio retargeting, the reference and test contents lose pixel correspondences. In dynamic range retargeting, the reference content is in HDR whereas the test contents are in SDR. Such differences between the reference and retargeted content prevent us from directly using the quality assessment techniques intended for non-retargeted content and forces us to rethink the way we perform quality assessment in the presence of retargeted content.

It turns out that extending existing quality assessment methods for non-retargeted content is possible in certain cases. In the remainder of this section we discuss a case study for quality assessment of tone retargeted (tone-mapped) images. An extensive review of current tone-mapping techniques was presented in Section 2.7 of this book.

6.1   Problem Definition: What Is Objective Quality Assessment?

In the most general sense, the goal of objective quality assessment is obtaining a quantitative measure that is, in the ideal case, highly correlated with the perceived quality of the input visual content. In objective quality assessment, the task of predicting such a measure is fulfilled by a computational *quality metric*. The exact quality assessment problem definition depends on multiple factors, including the availability of various input types, the variety of visual degradations assumed to influence quality, the overall philosophy of the metric design, and the level of visual system mechanisms being considered. In this section, we briefly compare these various approaches of objective quality assessment. We will also briefly discuss subjective methods in the next section.

Perhaps the most limiting factor that influences quality metric design is the availability of the reference content. While all quality metrics require the content whose quality is being assessed (namely the test content), some metrics also assume that corresponding reference content that has no visual degradations, and hence has perfect quality, is also given as input. For example, in case of assessing video coding artifacts the reference content is typically an uncompressed video. Similarly, in the quality assessment of computer generated imagery, the reference content could be a scene generated with a very precise (but perhaps computationally expensive) global illumination method.

Full-reference quality metrics assume that the reference content is available. Under this assumption, the quality assessment problem can be reduced to the assessment of the fidelity to the reference content. Therefore the predictions of full-reference metrics generally correlate better with subjective data compared to other metric types. On the downside, one can easily imagine practical scenarios where the required reference content is not available, where one can make use of a *no-reference* metric. At the other end of the spectrum, no-reference metrics predict the quality of input content by analyzing only the test content. An in-between strategy, which is relevant in some practical situations, is *reduced-reference* quality assessment. In this case, while the reference content is still not fully available to the metric, a reduced set of the reference's information content (e.g., certain image features) are made available.

Quality metrics can also be classified in terms of their generality: *specialized* metrics often focus only on a single type of distortion (e.g., noise, blockiness, etc.), whereas *general-purpose* metrics can predict the effect of any visual degradation on the perceived quality. Full-reference general-purpose objective quality assessment metrics often focus on predicting the visible differences between the test and reference contents to estimate the perceived quality of the test content.

In terms of modeling philosophy, *bottom-up* metrics seek to faithfully model the multitude of individual components that affect the perceived visual quality in the hope that together, those components will be able to make precise predictions. This approach usually results in algorithmically complex and computationally expensive metrics. On the other hand, *top-down* approaches model the perception of quality based on high-level assumptions on how the HVS is supposed to work, which may not always correspond to how it actually works. Generally speaking, there is only a thin line that separates bottom-up and top-down approaches. It is sometimes hard to discriminate between rigorous modeling of physiology and high-level assumptions, due to our limited understanding of the physiology of the HVS and the limits of current models of the various visual system mechanisms.

One can also classify quality metrics in terms of the level of visual system mechanisms being considered. *Low-level* quality metrics take into account mechanisms of low-level vision, such as luminance adaptation, contrast sensitivity, and masking. On the other hand, another line of metrics focuses on higher-level cognitive processes such as the perception of sharpness, clarity, and scene composition and try to predict visual quality in terms of aesthetics.

In a perfect world, we would have a no-reference, general-purpose metric that comprises an exact bottom-up visual system model, and therefore could predict both low, and high-level aspects of quality. However due to the enormous complexity of the HVS and our partial understanding of its working principles, such an ideal metric is not on the horizon yet. All the classifications mentioned in this section essentially describe the different compromises that are made to build quality metrics that consider a subset of the whole quality assessment problem.

6.2   Objective vs. Subjective Quality Assessment

Compared to objective quality assessment performed through computational metrics, subjective experiments offer a significantly more accurate alternative for measuring perceived quality. In visual computing literature, two main types of subjective studies are employed, namely, *rating* experiments, and *pairwise comparison* experiments. Figure 6.1 shows a brief illustration of these two methodologies. In this section, we will briefly discuss subjective quality assessment and point out its advantages and disadvantages with respect to objective quality assessment.

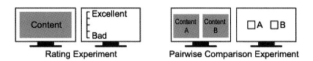

Figure 6.1. A comparison of experimental procedures. See text for details.

Two common protocols for rating studies are the single stimulus and double stimulus experiments. In both protocols, for each trial the subject is asked to provide a quality rating of the presented visual content. The scale of this rating is often discretized to five points ranging from excellent to bad. A minor difference between the single and double stimulus protocols is that in the former, only a single stimulus (test) is presented at each trial, whereas in the latter, two stimuli (test and reference) are presented side by side. The main difficulty with rating studies is the overall difficulty of judging absolute quality levels for human subjects. While we can easily spot any visible differences between two visual stimuli, quality ratings without the presence of any reference could easily be biased by each subject's personal preferences. Therefore, the reliability of subjects' responses tend to be low in rating studies, which as a consequence require more subjects and/or repetitions to obtain statistically significant outcomes.

In contrast, humans are much better in making comparative judgments where they are asked to choose the higher-quality one among two stimuli, instead of rating each stimulus on a (five-point) scale without any reference. Consequently, pairwise comparison protocols in general take less time per trial and require less repetitions to obtain statistically significant results compared to rating studies. The main problem with pairwise comparison studies is the sheer number of trials that need to be performed by each subject. In a naïvely implemented pairwise comparison study, each subject has to perform $\frac{n(n-1)}{2}$ trials, where n is the number of different stimuli (compared to n trials in a rating study). However, by utilizing the transitive relation between various stimuli and efficient sorting algorithms, the number of trials for pairwise comparison studies can be significantly reduced to approximately $n \, logn$ [180].

The major practical limitations of subjective experiments stem from the human involvement in the process. The design, setup, and execution of a subjective study often requires a notable amount of resources and time investment. The specific skill set and experience for performing subjective experiments may not always be present within a visual computing–oriented research and/or development team. When these limitations make subjective experiments infeasible, often the only option is to rely on objective metrics that would perform the task without requiring human involvement, at the cost of sacrificing quality prediction accuracy.

Another practically relevant example where subjective methods cannot be employed is when the quality prediction is used in a larger computational system. For example, in an iterative global illumination workflow, a quality metric can be employed for determining by how much the rendered image from the current iteration improved over the previous iteration. Hence, the iterative process can be terminated if the quality improvement between consecutive iterations falls below a certain threshold. In such a system one also has to rely on objective quality metrics, as running a subjective experiment between the iterations of a rendering algorithm is not a viable option. Similar arguments can also be made for real-time applications such as broadcasting.

In the remainder of this chapter we will focus on objective quality assessment metrics. In the next section we continue our discussion with the advantages and disadvantages of utilizing HVS models in objective quality assessment.

6.3 Human Visual System–Based Quality Assessment Metrics

As we discussed earlier in this chapter, no objective quality metric can perfectly predict visual quality at every instance. Each metric in the literature makes compromises in different areas. For a given task, the criteria for choosing the right metric among the multitude of alternatives often highly depends on practical constraints imposed by the task at hand. For example, the image and video quality community has been using a number of *mathematically based* quality metrics for predicting the effect of compression artifacts on visual quality. Among those, perhaps the simplest quality metric of all is the *mean squared error (*MSE*)*:

$$\text{MSE(I, J)} = \sqrt{\frac{1}{N} \sum_{i=1}^{N} (\text{I}_i - \text{J}_i)^2}, \qquad (6.1)$$

where I and J are gamma-corrected input images both having N pixels. For video sequences, first the metric outcome is computed for each frame individually, and then these individual per-frame quality predictions are averaged over the entire video sequence.

In practice, it is more common to use a scaled version of MSE, formulated as follows:

$$\text{PSNR(I, J)} = 20 \; log_{10} \frac{D}{\text{MSE(I, J)}}, \qquad (6.2)$$

where D denotes the dynamic range parameter (255 for 8-bit images). As a result of PSNR being widely used by the image and video quality community in numerous experiments, researchers working in this field have developed an intuition for interpreting the PSNR scores. For example, when 8-bit images are concerned, 50 decibels (dB) is often considered as the threshold for *visual equivalence*, and 30–50 dB range is considered as acceptable quality.

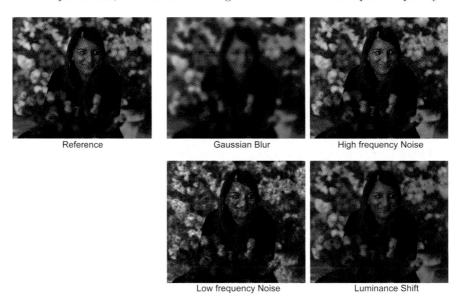

Figure 6.2. Reference image degraded with four different types of distortions. The PSNR value of each degraded image with respect to the reference image is approximately equal to 25 dB.

When PSNR and other similar mathematically based metrics are concerned, it is very important not to forget that such limits and ranges are only meaningful in a set of similar images (i.e., 8-bit) degraded with the same type of distortions (e.g., compression artifacts) under the same pre-defined viewing conditions. In fact, Figure 6.2 presents various degraded images that all yield approximately 25 dB. In other words, according to PSNR all degraded images in Figure 6.2 have the same quality. Yet we do not perceive all four degraded images as having the same level of quality. For example, the luminance shift is barely noticeable, whereas the low-frequency noise distortion is highly disturbing.

Some limitations of PSNR, such as the unrealistically high sensitivity to global luminance shift, is addressed in a slightly more sophisticated *structural similarity index metric* (SSIM), which is defined as follows:

$$\text{SSIM}(\text{I}, \text{J}) = l\left(\mu_\text{I}, \mu_\text{J}\right)^{\alpha} \; c\left(\sigma_\text{I}, \sigma_\text{J}\right)^{\beta} \; s\left(\sigma_\text{I}, \sigma_\text{J}\right)^{\gamma}, \tag{6.3}$$

where l, c, and s denote the luminance, contrast, and structure terms, μ and σ denote the mean and standard deviation, and α, β, and γ are model parameters that control the weights of the three terms of the metric. Thanks to its formulation, which treats luminance distortions separately from contrast and structural distortions, SSIM predicts that the luminance shift degradation in Figure 6.2 has very little effect on image quality (SSIM index $= 0.98$).

Despite the additional level of sophistication, SSIM still suffers from an important problem that is common in all mathematically based metrics—its outcome is scaled arbitrarily. Given two images degraded with the same type of distortion with different magnitudes, we can most likely determine which image is more degraded by comparing the corresponding SSIM indices. However, we won't be able to tell the severity of any degradation from the actual value of the SSIM index. For example, for distortion type A, a certain SSIM index value might correspond to a test image that is visually equivalent to the reference, whereas for another distortion type B, the same SSIM index might correspond to a test image with visible distortions.

The goal of *HVS-based metrics* is ensuring that the metric's outcome is scaled in perceptually uniform units. That is, instead of reporting a number arbitrarily within $[-1, 1]$ or in dB units, such metrics would report the probability of an average human observer detecting a difference between the input image pair. This type of output is very useful in practice where one would like to ensure that two images are visually equivalent. On the downside, however, predicting the average human observer's perception requires modeling various components of the human visual system. As a result, HVS-based metrics are algorithmically more complex than the mathematically based metrics, and often consume more computational resources.

We started discussing some fundamental characteristics of the HVS in Chapter 1 of this book. In the next paragraphs, we briefly discuss some additional HVS mechanisms that play an important role in the perception of visual quality.

It is well known that the human eye is not sensitive to absolute luminance levels. Whenever such a global luminance change occurs, our eyes adapt to this change gradually over a certain time period. For example, when driving out of a tunnel on a sunny day we are momentarily blinded, but then quickly recover our vision through *luminance adaptation*. This property of the HVS has been investigated in detail in the literature [18, 125]. The adaptation state of the viewer's visual system has a significant effect on his sensitivity to the degradations in the image quality, and consequently to her perception of image quality.

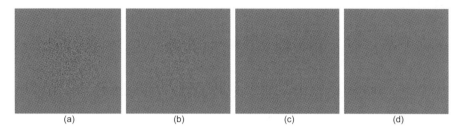

(a)   (b)   (c)   (d)

Figure 6.3. The spatial frequency of contrast strongly affects its visibility. Images a–d show the same pattern with the same magnitude but decreasing spatial frequency.

The sensitivity to degradations also highly depends on the spatial frequency of the distortion signal. Figure 6.3 shows random noise at the same magnitude but at 4 different peak spatial frequencies, ranging from the highest (a) to the lowest (d). While at the highest spatial frequency (a) the distortion pattern is clearly visible, its visibility decreases gradually as we lower the spatial frequency. In fact, the HVS is tuned to be more sensitive to a certain range of spatial frequencies, which is often modeled through a *Contrast Sensitivity Function (CSF)*.

Another HVS mechanism that significantly influences the sensitivity to degradations is *visual masking*, which refers to the decreased visual sensitivity to visual artifacts that have similar spatial frequency and orientation to the content. Figure 6.4 shows an example where the reference image (a) is distorted with two horizontal and one vertical sinusoidal gratings. Since the image content is dominated by horizontal contrast, the vertical grating results in low visual masking (b). Similarly, the horizontal grating with a higher spatial frequency than the image content is highly visible (c). The final grating with matching frequency and orientation is much harder to detect due to strong visual masking (d).

A number of objective quality metrics model HVS mechanisms such as contrast sensitivity and visual masking, as well as other mechanisms discussed earlier in this book, in order to accurately predict the visibility of visual artifacts. In the next chapter, we continue our discussion with such an HVS-based metric, specialized for tone mapped-images.

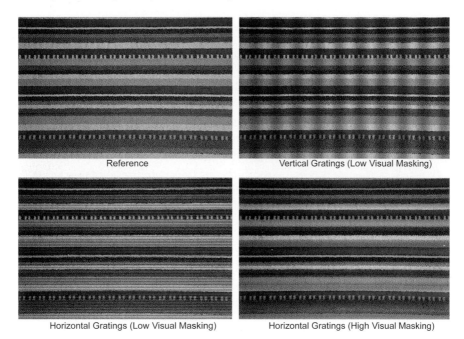

| Reference | Vertical Gratings (Low Visual Masking) |
| Horizontal Gratings (Low Visual Masking) | Horizontal Gratings (High Visual Masking) |

Figure 6.4. An illustration of visual masking. The content of the reference image is dominated by horizontally oriented contrast (a). Artificially added vertical gratings are immediately visible (b), similar to the horizontal gratings at a higher frequency than the image content (c). Despite having the same magnitude, the horizontal grating at the matching spatial frequency results in strong visual masking, and is therefore much harder to spot (d).

6.4 Dynamic Range Independent Quality Assessment

In this chapter, we present a full-reference human visual system–based metric for assessing the quality of luminance retargeted images. This metric is full-reference, thus we assume that both a reference image and a test image are available as input. In quality assessment of non-luminance-retargeted images, the high-level goal of a general-purpose quality metric is determining per-pixel visible differences between the two input images. Any such visible difference is then treated as a visual artifact that degrades image quality.

This process is rather straightforward if both images have the same dynamic range, since any deviation in the test image from the reference image can immediately be classified as a visual artifact. However, if the reference image is in HDR whereas the test image is in SDR, then the deviations from the reference image can either be genuine visual artifacts, or they can be due to the dynamic range difference between the input images. In the latter case, the visible differences in the test image should not be classified as visual artifacts. It is also easy to see that the same problem exists if the reference image is SDR and the test image is HDR, such as in inverse tone-mapping applications. The additional challenge for the quality assessment of luminance retargeted content is finding a new quality measure that is *dynamic range independent*, meaning that it can provide a meaningful estimate of image quality even when the reference and test images have different dynamic ranges.

The *Dynamic Range Independent Image Quality Metric (DRI-IQM)* [20] achieves this by processing both the reference and test images through an HVS model pipeline to obtain per-visual-channel human visual system responses scaled in *just noticeable difference (JND)* units, and then computing a set of structural distortions of the test image with respect to the reference image. The main processing steps of DRI-IQM are illustrated in Figure 6.5. In the remainder of this chapter, we will go through each of these steps and discuss them in detail.

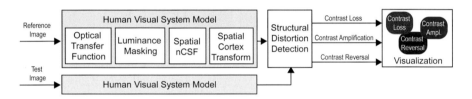

Figure 6.5. Major processing steps of the DRI-IQM metric. See text for details.

DRI-IQM comprises a relatively complex HVS model that is used to process both the reference and test images. While discussing DRI-IQM's HVS model we use L to refer to an input image's luminance. As Figure 6.5 illustrates, the same HVS model is applied to both the test and reference images, and therefore L denotes the luminance of either of the two input images.

The first component of the HVS model is the simulation of the effect of light scattering in the human eye using an *optical transfer function (OTF)*. In practice, this so-called *disability glare* effect is especially noticeable when looking at bright light sources in otherwise dim environments, i.e., looking at the headlights of a car at night, and manifests itself as reduced contrast sensitivity near the light source.

While highly sophisticated methods have been proposed for glare simulation (such as Ritschel et al. [233]), for the purposes of image quality assessment, a simpler model based on frequency domain filtering is sufficient. In fact, DRI-IQM utilizes the OTF from Deeley et al. [66], which is defined as follows:

$$OTF = \exp\left[-\left(\frac{\rho}{20.9 - 2.1d}\right)^{1.3-0.07d}\right],\qquad(6.4)$$

and applied to the input image luminance L in the Fourier domain:

$$L_{OTF} = \mathscr{F}^{-1}\left\{\mathscr{F}\{L\} \cdot OTF\right\},\qquad(6.5)$$

where d is an eye pupil diameter in mm and ρ is spatial frequency in cycles per degree. The pupil diameter is calculated for a global adaptation luminance L_a using the formula of Moon and Spencer [188]:

$$d = 4.9 - 3\,tanh[0.4(\pi \log_{10} L_a - 0.5)].\qquad(6.6)$$

The global adaptation luminance L_a is computed as the geometric mean of the input luminance L.

The next component of DRI-IQM's HVS model is a model of *luminance masking*, which accounts for the nonlinearity of luminance perception. To that end, DRI-IQM utilizes the JND-space transformation, where the input luminance L is mapped to perceptual JND units using a transfer function constructed from the peak detection thresholds [173]. One can construct such a transfer function by populating a vector T using the following recursive formula:

$$T[i] = T[i-1] + cvi(T[i-1])\,T[i-1]\quad \text{for } i = 2..N,\qquad(6.7)$$

where $T[1]$ is the minimum luminance we want to consider ($10^{-5}\ cd/m^2$ in case of DRI-IQM). The vector T serves as a lookup table for mapping the input luminance processed by the OTF (L_{OTF}) to the photoreceptor response R. The following step of DRI-IQM's HVS model uses the photoreceptor response R as its input. The mapping from luminance to JND values is performed simply by linear interpolation between the nearest pair of i values corresponding to the luminance of each pixel in L_{OTF}. The *contrast versus intensity (cvi)* function used in the recursive formula above estimates the lowest detection threshold at a particular adaptation level:

$$cvi(L_a) = (max_\rho\,[CSF(\rho, L_a, \cdot)])^{-1},\qquad(6.8)$$

where CSF is the static *contrast sensitivity function*, ρ denotes spatial frequency, and L_a denotes the adaptation luminance. The JND-space formulation makes the conservative assumption that the human eye can perfectly adapt to the area of a single image pixel, and therefore the input image's luminance is used as the adaptation map ($L_a = L$).

DRI-IQM utilizes the CSF proposed by Daly [60][1]:

$$CSF(\rho, L_a, \theta, i^2, d, c) = P \cdot \min \left[S_1 \left(\frac{\rho}{r_a \cdot r_c \cdot r_\theta} \right), S_1(\rho) \right], \qquad (6.9)$$

where

$$
\begin{aligned}
r_a &= && 0.856 \cdot d^{0.14} \\
r_c &= && \tfrac{1}{1+0.24c} \\
r_\theta &= && 0.11 \cos(4\theta) + 0.11 \\
S_1(\rho) &= && \left[\left(3.23(\rho^2 i^2)^{-0.3} \right) \right)^5 + 1 \right]^{-\frac{1}{5}} \cdot \\
& && A_l \epsilon \rho e^{-(B_l \epsilon \rho)} \sqrt{1 + 0.06 e^{B_l \epsilon \rho}} \\
A_l &= && 0.801 \left(1 + 0.7\, L_a^{-1} \right)^{-0.2} \\
B_l &= && 0.3 \left(1 + 100\, L_a^{-1} \right)^{-0.15}.
\end{aligned}
\qquad (6.10)
$$

The parameters of the CSF are as follows: ρ denotes the spatial frequency in cycles per visual degree, L_a denotes light adaptation level in cd/m^2, θ denotes orientation, i^2 denotes stimulus size in deg^2 ($i^2 = 1$), d denotes viewing distance in meters, c denotes eccentricity ($c = 0$), and ϵ is a model constant ($\epsilon = 0.9$). While one may be intimidated by the number of parameters of Daly's CSF, DRI-IQM uses the default values (given in parentheses) for many of them, which reduces the actual degrees of freedom to four.

Once we compute the per-pixel photoreceptor response R corresponding to the input image's luminance L, the next step in DRI-IQM's human visual system model is the application of the *neural* contrast sensitivity function (nCSF), which accounts for the spatial frequency and orientation dependency of contrast sensitivity. Different from the CSF (Equation 6.9), which takes input the image's per-pixel luminance as input and scales the input luminance in detection threshold units, $nCSF$ takes as input the per-pixel luminance already scaled in JND units. The sole purpose of $nCSF$ is to scale the input JND values, such that they also reflect the spatial frequency dependence of the human visual system's sensitivity.

In DRI-IQM, $nCSF$ is derived by removing the full CSF (Equation 6.9) JND scaling and the effect of disability glare:

$$nCSF(\rho, \theta, L_{la}, i^2, d, c) = \frac{CSF(\rho, \theta, L_a, i^2, d, c) \cdot cvi(L_a)}{OTF(\rho)}, \qquad (6.11)$$

and is applied to the photoreceptor response R as a filter in the frequency domain as follows:

$$R_{CSF} = \mathscr{F}^{-1} \left\{ \mathscr{F}\{R\} \cdot nCSF \right\}. \qquad (6.12)$$

---

[1] Note that the formulas for A_l and B_l contain corrections found after the correspondence with the author of the original publication.

Note that prior to filtering with the $nCSF$, the photoreceptor response R already accounts for disability glare (Equation 6.5) and luminance masking 6.7. With the application of $nCSF$, the photoreceptor response R is also modulated to model the visual system's sensitivity variation due to spatial frequency selectivity. In practice, since the $nCSF$ filter function depends on the local luminance of adaptation, the same kernel cannot be used for the entire image. To speed up computations, the photoreceptor response R is filtered six times assuming $L_a = \{\ 0.001,\ 0.01,\ 0.1,\ 1,\ 10,\ 100\ \}\ cd/m^2$ and the final value for each pixel is found by the linear interpolation between the two filtered maps closest to the L_a for a given pixel.

The final component of DRI-IQM's human visual system model is the *cortex transform*, which decomposes the photoreceptor response R into a number of spatial frequency- and orientation-bands. As such, the cortex transform models the behavior of frequency and orientation selective neurons in the primary visual cortex. Figure 6.6 illustrates how the cortex transform bands are arranged in the Fourier domain, and shows the inverse Fourier transform of selected cortex bands of an example image. Note that the center band is a low-pass filter, whereas all the other bands are band-pass filters.

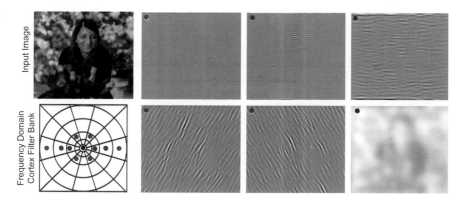

Figure 6.6. Cortex transform is formulated as a frequency domain filter bank. This figure shows selected cortex bands of the input image (that are transformed back to the spatial domain). Note the different spatial frequencies and orientations of the contrast patterns in the cortex bands.

In DRI-IQM, the cortex transform is formulated as a collection of the band-pass and orientation-selective filters [60, 287]. The band-pass filters are computed as

$$dom_k = \begin{cases} mesa_{k-1} - mesa_k & \text{for } k = 1..K - 2 \\ mesa_{k-1} - base & \text{for } k = K - 1 \end{cases} \qquad (6.13)$$

where K is the total number of spatial bands and the low-pass filters $mesa_k$ and $baseband$ have the form:

$$mesa_k = \begin{cases} 1 & \text{for } \rho \leq r - \frac{tw}{2} \\ \frac{1}{2}\left(1 + \cos\left(\frac{\pi(\rho - r + \frac{tw}{2})}{tw}\right)\right) & \text{for } r - \frac{tw}{2} < \rho \leq r + \frac{tw}{2} \\ 0 & \text{for } \rho > r + \frac{tw}{2} \end{cases}$$

$$base = \begin{cases} e^{-\frac{\rho^2}{2\sigma^2}} & \text{for } \rho < r_{K-1} + \frac{tw}{2} \\ 0 & \text{otherwise,} \end{cases}$$

(6.14)

where

$$r = 2^{-k}, \quad \sigma = \frac{1}{3}\left(r_{K-1} + \frac{tw}{2}\right) \quad \text{and} \quad tw = \frac{2}{3}r. \tag{6.15}$$

The orientation-selective filters are defined as

$$fan_l = \begin{cases} \frac{1}{2}\left(1 + \cos\left(\frac{\pi|\theta - \theta_c(l)|}{\theta_{tw}}\right)\right) & \text{for } |\theta - \theta_c(l)| \leq \theta_{tw} \\ 0 & \text{otherwise,} \end{cases} \tag{6.16}$$

where $\theta_c(l)$ is the orientation of the center, $\theta_c(l) = (l - 1) \cdot \theta_{tw} - 90$, and θ_{tw} is the transitional width, $\theta_{tw} = 180/L$. The cortex filter is formed by the product of the dom and fan filters:

$$B^{k,l} = \begin{cases} dom_k \cdot fan_l & \text{for } k = 1..K - 1 \text{ and } l = 1..L \\ base & \text{for } k = K. \end{cases} \tag{6.17}$$

Once the cortex transform's filter bank is generated according to Equation 6.17, each cortex band is generated through filtering in the Fourier domain as follows:

$$R^{k,l}_{cortex} = \mathscr{F}^{-1}\left\{\mathscr{F}\{R_{CSF}\} \cdot B^{k,l}\right\}. \tag{6.18}$$

The application of the cortex transform concludes the series of operations comprising DRI-IQM's HVS model. This cascade of operations transforms an input image's luminance to 31 cortex bands, which are scaled in perceptual JND units and accounts for disability glare, luminance masking, and the frequency dependency of contrast sensitivity.

As we mentioned in the beginning of this section, DRI-IQM processes both the reference and test images through the whole human visual system model and obtains two sets of cortex bands. At this point, general-purpose image quality assessment metrics such as HDR-VDP would compute the differences between the corresponding cortex bands of the reference and test images. These differences would be combined using a probability summation and a visible differences map is obtained. The detection probability of the normalized contrast response at each cortex band is computed using the *psychometric function*

$$P(R_{cortex}^{k,l}) = 1 - \exp(-|R_{cortex}^{k,l}|^3). \qquad (6.19)$$

In case of visible differences metrics, the distortion map resulting from the probability summation shows the per-pixel detection probability of the visible differences between the reference and test images.

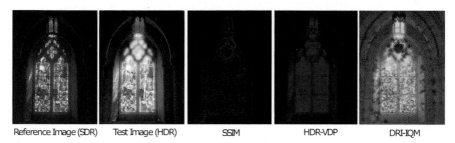

Reference Image (SDR)   Test Image (HDR)   SSIM   HDR-VDP   DRI-IQM

Figure 6.7. The outcome of SSIM, HDR-VDP, and DRI-IQM given the reference and test images. The visible difference maps of SSIM and HDR-VDP color-code the detection probability, where red/magenta indicates high probability, and green/yellow indicates low probability. The distortion map of DRI-IQM, on the other hand, color-codes contrast loss with green, contrast amplification with blue, and contrast reversal with red. Note that the visible difference maps of both SSIM and HDR-VDP are completely saturated and thus do not convey any useful information. DRI-IQM, on the other hand, detects three types of distortions, which shows both the effect of the input images having different dynamic ranges (contrast amplification shown in blue) and the effect of the local Gaussian blur artifact (contrast loss shown in green and contrast reversal shown in red).

Unfortunately the standard visible differences approach does not work when we seek to compare a pair of images with different dynamic ranges. Figure 6.7 shows a toy example that illustrates the problem. In this case, an HDR test image is being compared with an SDR reference image. Note that, in addition to the dynamic range difference, the test image also has a local Gaussian blur artifact. A comparison of the outcomes of two visible difference metrics SSIM and HDR-VDP with the outcome of DRI-IQM shows that the visible difference maps of both SSIM and HDR-VDP are almost entirely saturated simply because of the visible differences due to the different dynamic ranges of the input image pair.

This tells us that there is a visible difference between every corresponding pixel of the test and reference images, which will be noticed by an average human observer with very high probability. This statement is actually correct: There are, in fact, significant visible differences between the SDR reference and HDR test image. While being correct, predicting this difference does not give us any useful information in this case. What we rather would know is in how the test image is visually different compared to the reference image.

Note that there is also no indication of the local Gaussian blur artifact of the test image in the visible difference maps of SSIM and HDR-VDP in Figure 6.7. Since these metrics do not differentiate between the different *types* of visible differences, their outcomes can be confusing in cases where the test image has multiple sources of visible differences.

The main innovation of DRI-IQM, which enables a meaningful comparison of tone-mapped images, is the detection of three different per-pixel distortion types:

- **Loss of visible contrast** happens when visible contrast at some pixel of the reference image is not visible at the corresponding pixel of the test image.

- **Amplification of invisible contrast** happens when invisible contrast at some pixel of the reference image is amplified beyond the visibility threshold at the corresponding pixel of the test image.

- **Reversal of visible contrast** happens when positive (negative) visible contrast at some pixel of the reference image becomes negative (positive) visible contrast at the corresponding pixel of the test image.

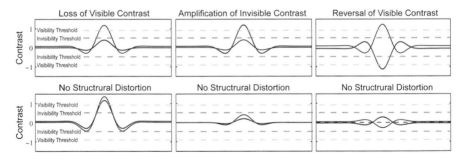

Figure 6.8. Illustration of the structural distortion detection mechanism of DRI-IQM. The top row shows cases that DRI-IQM detects as structural distortions, whereas the bottom row shows cases where no distortions are detected.

These three distortion types, which are specifically designed to handle input image pairs with different dynamic ranges, are illustrated in Figure 6.8. Their advantage for the quality assessment of tone mapping is illustrated in Figure 6.7. Different than SSIM and HDR-VDP, the distortion map of DRI-IQM indicates that the HDR test image has higher contrast in most regions compared to the SDR reference. DRI-IQM's outcome also clearly identifies the local Gaussian blur artifact, which causes contrast loss and contrast reversal in the test image.

We can formally express these three types of distortions separately for each band as the following conditional probabilities:

$$
\begin{aligned}
\text{loss of visible contrast:} & \quad P_{loss}^{k,l} = P_{r/v}^{k,l} \cdot P_{t/i}^{k,l}, \\
\text{amplification of invisible contrast:} & \quad P_{ampl}^{k,l} = P_{r/i}^{k,l} \cdot P_{t/v}^{k,l}, \\
\text{and reversal of visible contrast:} & \quad P_{rev}^{k,l} = P_{r/v}^{k,l} \cdot P_{t/v}^{k,l} \cdot R^{k,l},
\end{aligned}
\tag{6.20}
$$

where k and l are the spatial band and orientation indices, the subscript $r/\cdot$ denotes the reference and $t/\cdot$ the test image, the subscript $\cdot/v$ the visible and $\cdot/i$ the invisible contrast.

The probabilities $P_{\cdot/v}$ and $P_{\cdot/i}$ are found from the detection probabilities. HVS models commonly assume that a contrast is visible when it is detectable as in the two alternative forced choice (2AFC) experiments. This assumption has been found to be too conservative, since complex images are never scrutinized as much as the stimuli in such experiments. Therefore they require contrast to be detected with a higher probability of being regarded as visible. From an empirical study on a series of simplified stimuli, it was found that a reliable predictor of visible contrast is given by shifting the psychometric function, so that a contrast magnitude is *visible* with 50% probability, if it can be *detected* by our predictor with 95% probability (about 2 JND). The probability of invisible contrast is given by the negation of the probability of detection.

The rules from Equation 6.20 contain nonlinear operators, therefore the resulting probability map $P^{k,l}$ can contain features of spatial frequency that do not belong to a particular band. This leads to spurious distortions. To avoid this problem, each probability map is once more filtered using the corresponding cortex filter $B^{k,l}$:

$$
\hat{P}_{loss}^{k,l} = \mathscr{F}^{-1}\left\{ \mathscr{F}\{P_{loss}^{k,l}\} \cdot B^{k,l} \right\},
\tag{6.21}
$$

where $\mathscr{F}$ and $\mathscr{F}^{-1}$ are the 2D Fourier and inverse Fourier transforms, and the formula for $B^{k,l}$ is given in Equation 6.17. Assuming that detection of each distortion in each band is an independent process, the probability that a distortion will be detected in any band is given by

$$
P_{loss} = 1 - \prod_{k=1}^{N} \prod_{l=1}^{M} \left(1 - \hat{P}_{loss}^{k,l} \right).
\tag{6.22}
$$

The probability maps P_{ampl} and P_{rev} are computed in a similar way.

Unlike typical HVS-based visible difference predictors, DRI-IQM does not model visual masking (decrease in sensitivity with increase of contrast amplitude). Since DRI-IQM is invariant to suprathreshold contrast modifications, visual masking does not affect its result. The contrast difference is not relevant for DRI-IQM, therefore there is no need to estimate the magnitude of suprathreshold contrast in JND units.

DRI-IQM enables the objective quality assessment of dynamic range retargeted images. Figure 6.9 shows tone-mapped versions of an input HDR image using five well-known tone-mapping operators from the literature. DRI-IQM correctly detects the varying degrees of contrast loss in Drago [72], Durand and Dorsey [74], Pattanaik [213], and Reinhard [229] operators. It also detects the highly visible local contrast amplification in Fattal's [86] operator. Note that most operators result in varying degrees of contrast reversal, which could indicate the well-known *halo* artifacts commonly encountered in tone mapping.

DRI-IQM's distortion maps provide an objective basis for evaluating tone-mapped images with respect to the corresponding HDR references. Creative processes such as tone mapping can have multiple equally valid goals. For example, some tone-mapping algorithms seek to produce an SDR image that looks as close as possible to how the human eye perceives the reference HDR images. On the other hand, some tone-mapping operators mimic common photographic practices with the goal of achieving a photographic look. In comparison, some other tone-mapping operators are designed to show as much detail as possible in the SDR image, even though the result may look unnatural or non-photographic.

Using DRI-IQM's automatically generated distortion maps, one can formulate a number of quality measures that depend on tone-mapping goals. For example, such a quality measure could encourage a high number of pixels without any visible distortions of any type, if one seeks the SDR image that looks as close as possible to the HDR image. On the other hand, photographic tone-mapping operators usually utilize an S-shaped curve that overexposes the highlights and underexposes dark image regions. Accordingly, a quality measure for evaluating photographic look would not penalize detail loss near highlights and dark image regions. Finally, if the tone-mapping goal is maximum detail reproduction in the SDR image, then one can simply encourage detail amplification while penalizing contrast reversal. The reason for penalizing contrast reversal is to limit the halo artifacts that tend to appear near high-contrast edges if local details are significantly amplified. Other quality measures can be derived similarly for different tone-mapping goals.

6.5 Summary

Quality assessment of retargeted content, either subjective or objective, is an interesting and relatively unexplored research direction. Each type of content retargeting discussed in this book, be it color, tone, or spatial, comes with its individual quality assessment challenges. In this chapter we presented a method for objective quality assessment of tone-mapped images as a case study. The DRI-IQM metric presented in this chapter borrows components from standard objective quality assessment metrics, which would fail when faced with an image pair with different dynamic ranges. In comparison, we saw that DRI-IQM enabled a meaningful quality assessment of tone-mapped content by introducing new distortion definitions, rather than relying on visible differences. The same strategy of building upon established objective quality assessment methods by introducing new key components and modifications has also enabled the objective quality assessment of retargeted video sequences [19]. Beyond dynamic range retargeting, academic work has been done on quality assessment of other retargeting types, such as spatial retargeting [118, 162, 236]. With its broad scope and all its additional challenges over conventional quality assessment, the retargeting quality assessment will likely remain as an interesting field of study in the future.

Figure 6.9. The distortion maps obtained by comparing various tone-mapped images with the reference HDR image uncovers various characteristics of the tone-mapping methods used to generate them. For example, Pattanaik and Durand and Dorsey tend to lose details near the books, whereas Drago and Reinhard operators result in detail loss at highlights (green). The Fattal operator's result has significant detail amplification (blue) as shown by the corresponding distortion maps. All operators show some amount of contrast reversal (red). Using this automatically generated information, one can define rules for rating any tone-mapping operators. For example, one can derive a score from the number of pixels without any of the three distortions, or choose to penalize detail loss in certain image regions, etc.

A

Edge-Aware Filtering

In signal and image processing, linear filters are important tools for cutting

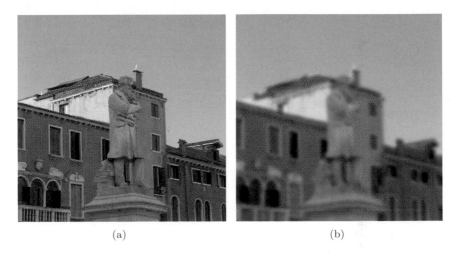

(a)   (b)

Figure A.1. An example of filtering with a Gaussian filter: (a) The input image. (b) The image in (a) filterd with $\sigma = 8$.

off certain unwanted frequencies and attenuating signals. Discrete low-pass filters applied to an image, I, can be defined as

$$F[I](\mathbf{x}) = \frac{\sum_{i \in \Omega(\mathbf{x})} f(\mathbf{x}, \mathbf{x}_i) I(\mathbf{x}_i)}{\sum_{i \in \Omega(\mathbf{x})} f(\mathbf{x}, \mathbf{x}_i)}, \qquad (A.1)$$

where $\mathbf{x}$ are the current pixel coordinates, Ω is the neighborhood around $\mathbf{x}$, and f is a discrete kernel representing the filter that has to be applied to I.

Typically, box, $f(\mathbf{x}, \mathbf{y}) = 1$, and Gaussian, $f(\mathbf{x}, \mathbf{y}) = \exp(-\|\mathbf{x} - \mathbf{y}\|^2 / 2\sigma^2)$, filters are employed in many image processing techniques for smoothing the image signal, e.g., removing noise. However, these techniques smooth not only details at small scales, which may be noise, but they also remove details at higher scales; i.e., strong edges. Figure A.1 shows a Gaussian filter applied to an image; the result of filtering produces a removal of part of the content of the image which results in blurring.

To avoid these issues, researchers have started to investigate edge-aware filters that try to apply a linear filter to the image while keeping *strong edges*. The number of edges to keep is typically specified by the user as a parameter. Typically, an edge-aware filter smooths an input image, I, while preserving *strong* edges. A seminal work in edge-aware filters is anisotropic

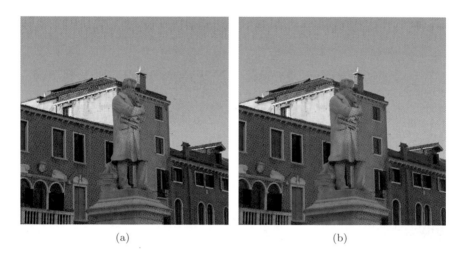

(a)   (b)

Figure A.2. An example of filtering with anisotropic diffusion applied to Figure A.1 (a): (a) Anisotropic diffusion after two iterations $K = 0.05$ and Equation A.3. Note that a few iterations are not enough to achieve smoothing, the filtered image is still close to the original one. (b) Anisotropic diffusion after 40 iterations $K = 0.05$ and Equation A.3; the image is now smoothed while preserving strong edges as defined by K.

diffusion [217], which diffuses color values to contiguous homogeneous regions except edges. This is achieved by using a modified version of the heat equation, which is defined as

$$\frac{\partial I}{\partial t} = \nabla g \cdot \nabla I + c(\mathbf{x}, t)\Delta t \quad c(\mathbf{x}, t) = g(\|\nabla I(\mathbf{x}, t)\|), \qquad (A.2)$$

where c is a flux function that controls the rate of diffusion, $\mathbf{x}$ are the pixel coordinates, and g is defined as

$$g(\|\nabla I(x, y, t)\|) = \exp\left(-\frac{\nabla I}{K}\right)^2 \quad \text{or} \tag{A.3}$$

$$g(\|\nabla I(x, y, t)\|) = \frac{1}{1 + \left(\frac{\|\nabla I\|}{K}\right)^2}, \tag{A.4}$$

where K is a parameter that defines when an edge can stop diffusion. While a large K value leads to isotropic diffusion (no edge-aware behavior); a small-medium K value preserves strong edges and diffuses flat regions. Note that g is a function of the image gradient, because it has to determine when to preserve it. Anisotropic diffusion, Equation A.2, has to be solved iteratively, and many iterations may be required, see Figure A.2. This may be an issue especially for fast filtering.

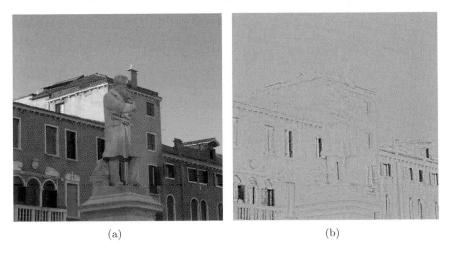

(a)   (b)

Figure A.3. An example of filtering with a bilateral filter applied to Figure A.1 (a): (a) The filtered image using two Gaussian functions for f_s and g_r which respectively have $\sigma_s = 8$ and $\sigma_r = 0.05$. This image is usually called the *base* layer in a bilateral decomposition. (b) The difference between Figure A.1 (a) and (a). This image is usually called the *detail* layer in a bilateral decomposition.

A method related to anisotropic diffusion [27, 210] is the bilateral filter [252, 268], a nonlinear filter that can be seen as an extension of classic linear filters, Equation A.1, where local image content, e.g., edges, is a weight. This is defined as

$$BF[I](\mathbf{x}) = \frac{\sum_{\mathbf{x}_i \in \Omega} I(\mathbf{x}_i) f_s(\|\mathbf{x} - \mathbf{x}_i\|) g_r(\|I(\mathbf{x}) - I(\mathbf{x}_i)\|)}{\sum_{\mathbf{x}_i \in \Omega} f_s(\|\mathbf{x} - \mathbf{x}_i\|) g_r(\|I(\mathbf{x}) - I(\mathbf{x}_i)\|)}, \tag{A.5}$$

where f_s and g_r are, respectively, the spatial and range function, which are typically Gaussian functions. However, other types of functions can also be used [74]. Ω is the neighborhood around the current pixel, and it is typically proportional to the kernel size of f_s. Furthermore, the filter can be used to decompose the image in a *base* layer, see Figure A.3 (a), and a *detail* layer, see Figure A.3 (b). This process can be also iterated for more scales.

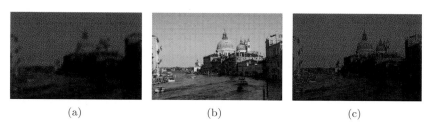

(a)   (b)   (c)

Figure A.4. An example of cross-bilateral filtering: (a) The input low-resolution image. (b) A guide image with target edges. (c) The result of the cross-bilateral filter where I is (a) and E is (b). Note that edges are successfully transferred.

This filter is very handy because it can be extended for transferring edges [75, 220] from a different image, E, even at a different resolution [143]. This variant of Equation A.5 is typically called a joint/cross-bilateral filter, and it is defined as

$$CBF[I, E](\mathbf{x}) = \frac{\sum_{\mathbf{x}_i \in \Omega} I(\mathbf{x}_i) f_s(\|\mathbf{x} - \mathbf{x}_i\|) g_r(\|E(\mathbf{x}) - E(\mathbf{x}_i)\|)}{\sum_{\mathbf{x}_i \in \Omega} f_s(\|\mathbf{x} - \mathbf{x}_i\|) g_r(\|E(\mathbf{x}) - E(\mathbf{x}_i)\|)}, \quad (A.6)$$

where E is an image with the edges to be transferred to I; see Figure A.4.

When Ω is large, Equation A.5 and Equation A.6 become computationally expensive. In order to efficiently evaluate them, several approximations [3, 4, 26, 49] have been presented, and they provide high-quality results for both SDR and HDR images. Moreover, this filter maintains edges but not other features such as ridges.

To reduce computational complexity and to maintain all features, many methods have been recently proposed based on different ideas such as wavelet [85], local statistics [137, 257], multi-scale pyramids [211], domain transform [92], adaptive manifolds [93], local linear model fitting [111], weighted least square [81], L0 gradient minimization [300], etc. All these methods provide edge preserving results, and they can be applied iteratively in a multi-scale fashion.

Bibliography

[1] Agoston G. A. *Color Theory and Its Application in Art Design*. Springer-Verlag, 1979.

[2] Arash Abadpour and Shohreh Kasaei. An efficient PCA-based color transfer method. *J. Vis. Comun. Image Represent.*, 18(1):15–34, February 2007.

[3] Andrew Adams, Jongmin Baek, and Myers Abraham Davis. Fast high-dimensional filtering using the permutohedral lattice. *Computer Graphics Forum*, 29(2):753–762, 2010.

[4] Andrew Adams, Natasha Gelfand, Jennifer Dolson, and Marc Levoy. Gaussian kd-trees for fast high-dimensional filtering. *ACM Trans. Graph.*, 28(3):1–12, 2009.

[5] Adobe Systems Inc. Photoshop CC, September 2015. http://www.adobe.com/products/photoshop.html.

[6] Ahmet Oğuz Akyüz, Roland Fleming, Bernhard E. Riecke, Erik Reinhard, and Heinrich H. Bülthoff. Do HDR displays support LDR content?: A psychophysical evaluation. *ACM Trans. Graph.*, 26(3):38, 2007.

[7] Ahmet Oğuz Akyüz and Erik Reinhard. Color appearance in high-dynamic-range imaging. *Journal of Electronic Imaging*, 15(3), July-September 2006.

[8] Xiaobo An and Fabio Pellacini. AppProp: All-pairs appearance-space edit propagation. *ACM Trans. Graph.*, 27(3):40:1–40:9, August 2008.

[9] Xiaobo An and Fabio Pellacini. User-controllable color transfer. *Computer Graphics Forum*, 29(2):263–271, May 2010.

[10] Codruta O. Ancuti, Cosmin Ancuti, Chris Hermans, and Philippe Bekaert. Image and video decolorization by fusion. In *Proceedings of the 10th Asian conference on Computer vision - Volume Part I*, ACCV'10, pages 79–92, Queenstown, New Zealand, 2011.

[11] Alessandro Artusi, Rafal Mantiuk, Thomas Richter, Pavel Korshunov, Philippe Hanhart, Touradj Ebrahimi, and Massimiliano Agostinelli. JPEG XT: A Compression Standard for HDR and WCG Images [Standards in a Nutshell]. *IEEE Signal Processing Magazine*, 33(2):118–124, 2016.

[12] Alessandro Artusi, Rafał Mantiuk, Richter Thomas, Hanhart Philippe, Korshunov Pavel, Agostinelli Massimiliano, Ten Arkady, and Ebrahimi Touradj. Overview and evaluation of the JPEG XT HDR image compression standard. *Real-Time Image Processing Journal*, pages 1–16, 2015.

[13] Alessandro Artusi, Zhuo Su, Zongwei Zhang, Dimitris Drikakis, and Xiaonan Luo. High-order wavelet reconstruction for multi-scale edge aware tone mapping. *Computers & Graphics*, 45:51–63, 2014.

[14] Alessandro Artusi and Alexander Wilkie. Color printer characterization using radial basis function networks. *Colour Imaging VI Device-Independent Colour, Colour Hardcopy, and Graphics Arts, SPIE, Electronic Imaging.*, 4300:70–80, 2001.

[15] Alessandro Artusi and Alexander Wilkie. Novel color printer characterization model. *Journal of Electronic Imaging*, 12(3):448–458, 2003.

[16] Michael Ashikhmin. A tone mapping algorithm for high contrast images. In *EGRW '02: Proceedings of the 13th Eurographics Workshop on Rendering*, pages 145–156, Aire-la-Ville, Switzerland, 2002. Eurographics Association.

[17] Shai Avidan and Ariel Shamir. Seam carving for content-aware image resizing. *ACM Trans. Graph.*, 26(3):10, 2007.

[18] T. O. Aydın, K. Myszkowski, and H-P. Seidel. Predicting display visibility under dynamically changing lighting conditions. *Computer Graphics Forum (Proc. of EUROGRAPHICS)*, 28(3), 2009.

[19] T. O. Aydın, M. Čadík, K. Myszkowski, and H-P. Seidel. Video quality assessment for computer graphics applications. In *To Appear in: ACM Transactions on Graphics (Proc. of SIGGRAPH Asia)*, 2010.

[20] Tunç Ozan Aydin, RafałMantiuk, Karol Myszkowski, and Hans-Peter Seidel. Dynamic range independent image quality assessment. *ACM Trans. Graph. (Proc. of SIGGRAPH)*, 27(3):69:1–69:10, August 2008.

[21] Soonmin Bae, Sylvain Paris, and Frédo Durand. Two-scale tone management for photographic look. In *ACM SIGGRAPH 2006 Papers*, SIGGRAPH '06, pages 637–645, New York, NY, USA, 2006. ACM.

[22] Raja Bala and Reiner Eschbach. Spatial color-to-grayscale transform preserving chrominance edge information. In *Color Imaging Conference*, pages 82–86. IS&T – The Society for Imaging Science and Technology, 2004.

[23] Francesco Banterle, Alessandro Artusi, Elena Sikudova, Thomas Edward William Bashford-Rogers, Patrick Ledda, Marina Bloj, and Alan Chalmers. Dynamic range compression by differential zone mapping based on psychophysical experiments. In *ACM Symposium on Applied Perception (SAP)*. ACM Press, August 2012.

[24] Francesco Banterle, Patrick Ledda, Kurt Debattista, and Alan Chalmers. Inverse tone mapping. In *GRAPHITE '06: Proceedings of the 4th International Conference on Computer Graphics and Interactive Techniques in Australasia and Southeast Asia*, pages 349–356, New York, NY, USA, 2006. ACM.

[25] Francesco Banterle, Patrick Ledda, Kurt Debattista, and Alan Chalmers. Expanding low dynamic range videos for high dynamic range applications. In *SCCG '08: Proceedings of the 4th Spring Conference on Computer Graphics*, pages 349–356, New York, NY, USA, 2008. ACM.

[26] Francesco Banterle and Roberto Scopigno. BoostHDR: A novel backward-compatible method for HDR images. In Andrew G. Tescher, editor, *Proceedings of SPIE Volume 8499: Applications of Digital Image Processing XXXV*, pages 3145–3148. SPIE, August 2012.

[27] Danny Barash. A fundamental relationship between bilateral filtering, adaptive smoothing, and the nonlinear diffusion equation. *IEEE Trans. Pattern Anal. Mach. Intell.*, 24(6):844–847, June 2002.

[28] Yoann Baveye, Fabrice Urban, Christel Chamaret, Vincent Demoulin, and Pierre Hellier. Saliency-guided consistent color harmonization. In *Proceedings of the 4th International Conference on Computational Color Imaging*, CCIW'13, pages 105–118, Berlin, Heidelberg, 2013. Springer-Verlag.

[29] R. S. Berns, R. J. Motta, and M. E. Gorzynski. CRT colorimetry part 1: Theory and practice. *Color Research and Application Journal*, 18(5):299–314, October 1993.

[30] R. S. Berns, R. J. Motta, and M. E. Gorzynski. CRT colorimetry part 2: Metrology. *Color Research and Application Journal*, 18(5):315–325, October 1993.

[31] M. Bertalmio, L. Vese, G. Sapiro, and S. Osher. Simultaneous structure and texture image inpainting. *Trans. Img. Proc.*, 12(8):882–889, August 2003.

[32] A. Bhattacharyya. On a measure of divergence between two multinomial populations. *The Indian Journal Of Statistics*, 7(4):401–406, 1946.

[33] Christian Bloch. *The HDRI Handbook 2.0: High Dynamic Range Imaging for Photographers and CG Artists*. Rocky Nook, January 2013.

[34] Jennifer Bonnard, Céline Loscos, Gilles Valette, Jean-Michel Nourrit, and Laurent Lucas. High-dynamic range video acquisition with a multiview camera. In *Photonics Europe 2012 : Optics, Photonics and Digital Technologies for Multimedia Applications*, Brussel, Belgium, April 2012. SPIE.

[35] Nicolas Bonneel, Kalyan Sunkavalli, Sylvain Paris, and Hanspeter Pfister. Example-based video color grading. *ACM Trans. Graph.*, 32(4):39:1–39:12, July 2013.

[36] Nicolas Bonnier, Francis Schmitt, Michael Hull, and Christophe Leynadier. Spatial and color adaptive gamut mapping: A mathematical framework and two new algorithms. *Proc. of the 15th Color Imaging Conference*, pages 267–272, 2007.

[37] Yuri Boykov, Olga Veksler, and Ramin Zabih. Fast approximate energy minimization via graph cuts. *IEEE Trans. Pattern Anal. Mach. Intell.*, 23(11):1222–1239, November 2001.

[38] Margarita Bratkova, Solomon Boulos, and Peter Shirley. oRGB: A practical opponent color space for computer graphics. *IEEE Comput. Graph. Appl.*, 29(1):42–55, January 2009.

[39] G. J. Braun, M. D. Fairchild, and F. Ebner. Color gamut mapping in a hue-linearized CIELAB color space. In *Proceedings of the 6th Color Imaging Conference. Scottsdale, AZ, USA*, pages 163–168, 1998.

[40] Matthew Brown and David G. Lowe. Automatic panoramic image stitching using invariant features. *Int. J. Comput. Vision*, 74(1):59–73, August 2007.

[41] Peter J. Burt and Edward H. Adelson. The Laplacian pyramid as a compact image code. In *Readings in Computer Vision: Issues, Problems, Principles, and Paradigms*, pages 671–679, San Francisco, CA, USA, 1987. Morgan Kaufmann Publishers Inc.

[42] Vladimir Bychkovsky, Sylvain Paris, Eric Chan, and Frédo Durand. Learning photographic global tonal adjustment with a database of input / output image pairs. In *The Twenty-Fourth IEEE Conference on Computer Vision and Pattern Recognition*, Colorado Springs, CO, June 2011.

[43] Martin Čadík. Perceptual evaluation of color-to-grayscale image conversions. *Comput. Graph. Forum*, 27(7):1745–1754, 2008.

[44] Susana Castillo, Tilke Judd, and Diego Gutierrez. Using eye-tracking to assess different image retargeting methods. In *Proceedings of the ACM SIGGRAPH Symposium on Applied Perception in Graphics and Visualization*, APGV '11, pages 7–14, New York, NY, USA, 2011. ACM.

[45] Che-Han Chang and Yung-Yu Chuang. A line-structure-preserving approach to image resizing. In *IEEE Conference on Computer Vision and Pattern Recognition (CVPR). Providence, RI, USA, June 16–21*, pages 1075–1082, June 2012.

[46] Huiwen Chang, Ohad Fried, Yiming Liu, Stephen DiVerdi, and Adam Finkelstein. Palette-based photo recoloring. *ACM Trans. Graph.*, 34(4):139:1–139:11, July 2015.

[47] Youngha Chang, Suguru Saito, and Masayuki Nakajima. A framework for transfer colors based on the basic color categories. In *2003 Computer Graphics International (CGI 2003), 9-11 July 2003, Tokyo, Japan*, pages 176–183, 2003.

[48] Youngha Chang, Suguru Saito, and Masayuki Nakajima. Example-based color transformation of image and video using basic color categories. *IEEE Transactions on Image Processing*, 16(2):329–336, 2007.

[49] Jiawen Chen, Sylvain Paris, and Frédo Durand. Real-time edge-aware image processing with the bilateral grid. *ACM Trans. Graph.*, 26(3):103, 2007.

[50] Renjie Chen, Daniel Freedman, Zachi Karni, Craig Gotsman, and Ligang Liu. Content-aware image resizing by quadratic programming. In *Proc. NORDIA*, 2010.

[51] Ming-Ming Cheng, Niloy J. Mitra, Xiaolei Huang, Philip H. S. Torr, and Shi-Min Hu. Global contrast based salient region detection. *IEEE TPAMI*, 37(3):569–582, 2015.

[52] Alex Yong-Sang Chia, Shaojie Zhuo, Raj Kumar Gupta, Yu-Wing Tai, Siu-Yeung Cho, Ping Tan, and Stephen Lin. Semantic colorization with internet images. *ACM Trans. Graph.*, 30(6):156:1–156:8, December 2011.

[53] Prasun Choudhury and Jack Tumblin. The trilateral filter for high contrast images and meshes. In *EGRW '03: Proceedings of the 14th Eurographics Workshop on Rendering*, pages 186–196, Aire-la-Ville, Switzerland, 2003. Eurographics Association.

[54] CIE. The CIE 1997 interim colour appearance model (simple version), CIECAM 97s. *Standard CIE TC1-34*, 1998.

[55] Daniel Cohen-Or, Olga Sorkine, Ran Gal, Tommer Leyvand, and Ying-Qing Xu. Color harmonization. *ACM Trans. Graph.*, 25(3):624–630, July 2006.

[56] Dorin Comaniciu and Peter Meer. Mean shift: A robust approach toward feature space analysis. *IEEE Trans. Pattern Anal. Mach. Intell.*, 24(5):603–619, May 2002.

[57] Corinna Cortes and Vladimir Vapnik. Support-vector networks. *Mach. Learn.*, 20(3):273–297, September 1995.

[58] Ming Cui, Jiuxiang Hu, Anshuman Razdan, and Peter Wonka. Color to gray conversion using ISOMAP. *The Visual Computer*, 26(11):1349–1360, 2010.

[59] Navneet Dalal and Bill Triggs. Histograms of oriented gradients for human detection. In *Proceedings of the IEEE Computer Society Conference on Computer Vision and Pattern Recognition (CVPR). San Diego, CA, USA, June 20–26*, volume 2, pages 886–893, June 2005.

[60] S. Daly. The Visible Differences Predictor: An algorithm for the assessment of image fidelity. In *Digital Images and Human Vision*, pages 179–206. MIT Press, 1993.

[61] Commission Internationale de l'Eclairage (CIE). Industrial colour-difference evaluation, 1995.

[62] Paul Debevec. A median cut algorithm for light probe sampling. In *SIGGRAPH '05: ACM SIGGRAPH 2005 Posters*, page 66, New York, NY, USA, 2005. ACM.

[63] Paul Debevec. HDRShop 3.0, December 2013. http://www.hdrshop.com.

[64] Paul Debevec and Jitendra Malik. Recovering high dynamic range radiance maps from photographs. In *SIGGRAPH '97: Proceedings of the 24th Annual Conference on Computer Graphics and Interactive Techniques*, pages 369–378, New York, NY, USA, 1997. ACM Press/Addison-Wesley Publishing Co.

[65] Doug DeCarlo and Anthony Santella. Stylization and abstraction of photographs. *ACM Trans. Graph.*, 21(3):769–776, July 2002.

[66] R. J. Deeley, N. Drasdo, and W. N. Charman. A simple parametric model of the human ocular modulation transfer function. *Ophthalmology and Physiological Optics*, 11:91–93, 1991.

[67] Julie Delon. Midway image equalization. *J. Math. Imaging Vis.*, 21(2):119–134, September 2004.

[68] Julie Delon, Agnès Desolneux, Jose Luis Lisani, and Ana Belén Petro. Automatic color palette. In *Proceedings of the 2005 International Conference on Image Processing (ICIP). Genoa, Italy, September 11–14*, pages 706–709, 2005.

[69] Silvano Di Zenzo. A note on the gradient of a multi-image. *Comput. Vision Graph. Image Process.*, 33(1):116–125, January 1986.

[70] Piotr Didyk, Rafał Mantiuk, Matthias Hein, and Hans-Peter Seidel. Enhancement of bright video features for HDR displays. *Computer Graphics Forum*, 27(4):1265–1274, 2008.

[71] Yi Dong and Dan Xu. Interactive local color transfer based on coupled map lattices. In *11th IEEE International Conference on Computer-Aided Design and Computer Graphics*, CAD/Graphics '09, pages 146–149, Washington, DC, USA, 2009. IEEE Computer Society.

[72] Frederic Drago, Karol Myszkowski, Thomas Annen, and Norishige Chiba. Adaptive logarithmic mapping for displaying high contrast scenes. *Computer Graphics Forum*, 22(3):419–426, 2003.

[73] Mark S. Drew, David Connah, Graham D. Finlayson, and Bloj Marina. Improved colour to greyscale via integrability correction. *Proc. SPIE 7240, Human Vision and Electronic Imaging XIV, 72401B*, 2009.

[74] Frédo Durand and Julie Dorsey. Fast bilateral filtering for the display of high-dynamic-range images. *ACM Trans. Graph.*, 21(3):257–266, 2002.

[75] Elmar Eisemann and Frédo Durand. Flash photography enhancement via intrinsic relighting. *ACM Trans. Graph.*, 23(3):673–678, 2004.

[76] Mark D. Fairchild and Garrett M. Johnson. The iCAM framework for image appearance, image differences, and image quality. *Journal of Electronic Imaging*, 13:126–138, 2004.

[77] Mark D. Fairchild and Elizabeth Pirrotta. Predicting the lightness of chromatic object colors using CIELAB. *Color Research and Application*, 16(16):385–393, 1991.

[78] M.D. Fairchild. *Color Appearance Models*. The Wiley-IS&T Series in Imaging Science and Technology. Wiley, 2005.

[79] Xin Fan, Xing Xie, He-Qin Zhou, and Wei-Ying Ma. Looking into video frames on small displays. In *Multimedia '03*, pages 247–250, New York, NY, USA, 2003. ACM.

[80] Zeev Farbman, Raanan Fattal, and Dani Lischinski. Diffusion maps for edge-aware image editing. *ACM Trans. Graph.*, 29(6):145:1–145:10, December 2010.

[81] Zeev Farbman, Raanan Fattal, Dani Lischinski, and Richard Szeliski. Edge-preserving decompositions for multi-scale tone and detail manipulation. *ACM Trans. Graph.*, 27(3):67:1–67:10, August 2008.

[82] Hany Farid. Blind inverse gamma correction. *IEEE Transactions on Image Processing*, 10(10):1428–1433, October 2001.

[83] H. S. Faridul, T. Pouli, C. Chamaret, J. Stauder, A. Tremeau, and E. Rein-
      hard. A survey of color mapping and its applications. In Sylvain Lefebvre
      and Michela Spagnuolo, editors, *Eurographics 2014 - State of the Art Reports.*
      The Eurographics Association, 2014.

[84] Hasan Sheikh Faridul, Jurgen Stauder, Jonathan Kervec, and Alain Trmeau.
      Approximate cross channel color mapping from sparse color correspondences.
      In *Proceedings of the 2013 IEEE International Conference on Computer
      Vision Workshops*, ICCVW '13, pages 860–867, Washington, DC, USA,
      2013. IEEE Computer Society.

[85] Raanan Fattal. Edge-avoiding wavelets and their applications. *ACM Trans.
      Graph.*, 28(3):22:1–22:10, July 2009.

[86] Raanan Fattal, Dani Lischinski, and Michael Werman. Gradient domain
      high dynamic range compression. *ACM Trans. Graph.*, 21(3):249–256, 2002.

[87] Jie Feng, Yichen Wei, Litian Tao, Chao Zhang, and Jian Sun. Salient
      object detection by composition. In *Proceedings of the 2011 International
      Conference on Computer Vision*, ICCV '11, pages 1028–1035, Washington,
      DC, USA, 2011. IEEE Computer Society.

[88] James A. Ferwerda, Sumanta N. Pattanaik, Peter Shirley, and Donald P.
      Greenberg. A model of visual adaptation for realistic image synthesis. In
      *Proceedings of the 23rd Annual Conference on Computer Graphics and
      Interactive Techniques*, SIGGRAPH '96, pages 249–258, New York, NY,
      USA, 1996. ACM.

[89] Hunt R. W. G. *Colorimetry.* Technical review. In C. Murchison (Ed.);
      Worchester, Massachusetts: Clark University Press., 1984.

[90] Ran Gal, Olga Sorkine, and Daniel Cohen-Or. Feature-aware texturing. In
      Eurographics Symposium on Rendering. Nicosia, Cyprus, June 26–28, pages
      297–303, 2006.

[91] Orazio Gallo, Marius Tico, Roberto Manduchi, Natasha Gelfand, and Kari
      Pulli. Metering for exposure stacks. *Computer Graphics Forum (Proceedings
      of Eurographics)*, 31:479–488, 2012.

[92] Eduardo S. L. Gastal and Manuel M. Oliveira. Domain transform for edge-
      aware image and video processing. *ACM Trans. Graph.*, 30(4):69:1–69:12,
      July 2011.

[93] Eduardo S. L. Gastal and Manuel M. Oliveira. Adaptive manifolds for
      real-time high-dimensional filtering. *ACM Trans. Graph.*, 31(4):33:1–33:13,
      July 2012.

[94] Arjan Gijsenij, Theo Gevers, and Joost van de Weijer. Improving color
      constancy by photometric edge weighting. *IEEE Trans. Pattern Anal. Mach.
      Intell.*, 34(5):918–929, 2012.

[95] Michael Goesele, Wolfgang Heidrich, and Hans peter Seidel. Color calibrated
      high dynamic range imaging with ICC profiles. In *In Proceedings of the
      9th Color Imaging Conference Color Science and Engineering: Systems,
      Technologies, Applications. Scottsdale, Arizona*, pages 286–290, 2001.

[96] Stas Goferman, Lihi Zelnik-Manor, and Ayellet Tal. Context-aware saliency detection. *Pattern Analysis and Machine Intelligence, IEEE Transactions on*, 34(10):1915–1926, October 2012.

[97] Dan B. Goldman. Vignette and exposure calibration and compensation. *IEEE Transactions on Pattern Analysis and Machine Intelligence*, 32(12):2276–2288, 2010.

[98] Rafael C. Gonzalez and Richard E. Woods. *Digital Image Processing (3rd Edition)*. Prentice-Hall, Inc., Upper Saddle River, NJ, USA, 2006.

[99] Amy A. Gooch, Sven C. Olsen, Jack Tumblin, and Bruce Gooch. Additional material of color2gray: Salience-preserving color removal. http://www.cs.northwestern.edu/~ago820/color2gray/.

[100] Amy A. Gooch, Sven C. Olsen, Jack Tumblin, and Bruce Gooch. Color2gray: Salience-preserving color removal. In *ACM SIGGRAPH 2005 Papers*, SIGGRAPH '05, pages 634–639, New York, NY, USA, 2005. ACM.

[101] Daniel Graf, Daniele Panozzo, and Olga Sorkine-Hornung. Mobile image retargeting. In *Proceedings of the Vision, Modeling and Visualization Workshop (VMV)*. Eurographics Association, 2013.

[102] Daniel Graf, Daniele Panozzo, and Olga Sorkine-Hornung. Mobile image retargeting. In *Proceedings of the Vision, Modeling and Visualization Workshop (VMV)*. Eurographics Association, 2013.

[103] Mark Grundland and Neil A. Dodgson. Decolorize: Fast, contrast enhancing, color to grayscale conversion. *Pattern Recognition*, 40(11):2891–2896, 2007.

[104] Chenlei Guo, Qi Ma, and Liming Zhang. Spatio-temporal saliency detection using phase spectrum of quaternion Fourier transform. In *IEEE Conference on Computer Vision and Pattern Recognition (CVPR). Anchorage, Alaska, USA, June 24–26*, pages 1–8, 2008.

[105] John Hable. Uncharted 2: HDR lighting. In *Game Developers Conference (GDC), San Francisco, California, March 9–13*, 2010.

[106] Yoav HaCohen, Eli Shechtman, Dan B. Goldman, and Dani Lischinski. Non-rigid dense correspondence with applications for image enhancement. *ACM Transactions on Graphics (Proceedings of ACM SIGGRAPH 2011)*, 30(4):70:1–70:9, 2011.

[107] Yoav HaCohen, Eli Shechtman, Dan B. Goldman, and Dani Lischinski. Optimizing color consistency in photo collections. *ACM Transactions on Graphics (Proceedings of ACM SIGGRAPH 2013)*, 32(4):85:1 – 85:9, 2013.

[108] Saghi Hajisharif, Jonas Unger, and Joel Kronander. HDR reconstruction for alternating gain (ISO) sensor readout. In *Proceedings of Eurographics Short Papers*, May 2014.

[109] C. Harris and M. Stephens. A combined corner and edge detector. In *Proceedings of the 4th Alvey Vision Conference. Manchester, UK*, pages 147–151, 1988.

[110] S. F. Hasan, J. Stauder, and A. Tremeau. Robust color correction for stereo. In *Proceedings of the 2011 Conference for Visual Media Production*, CVMP '11, pages 101–108, Washington, DC, USA, 2011. IEEE Computer Society.

[111] Kaiming He, Jian Sun, and Xiaoou Tang. Guided image filtering. In *Proceedings of the 11th European Conference on Computer Vision: Part I*, ECCV'10, pages 1–14, Berlin, Heidelberg, 2010. Springer-Verlag.

[112] Selig Hecht. *Vision II: The Nature of the Photoreceptor process. A Handbook of General Experimental Psychology*. Wiley Series in Pure and Applied Optics. In C. Murchison (Ed.); Worchester, Massachusetts: Clark University Press, 1934.

[113] Aaron Hertzmann, Charles E. Jacobs, Nuria Oliver, Brian Curless, and David H. Salesin. Image analogies. In *Proceedings of the 28th Annual Conference on Computer Graphics and Interactive Techniques*, SIGGRAPH '01, pages 327–340, New York, NY, USA, 2001. ACM.

[114] Derek Hoiem, Alexei A. Efros, and Martial Hebert. Recovering surface layout from an image. *Int. J. Comput. Vision*, 75(1):151–172, October 2007.

[115] Xiaodi Hou and Liqing Zhang. Color conceptualization. In *Proceedings of the 15th International Conference on Multimedia*, MULTIMEDIA '07, pages 265–268, New York, NY, USA, 2007. ACM.

[116] Xiaodi Hou and Liqing Zhang. Saliency detection: A spectral residual approach. In *IEEE Conference on Computer Vision and Pattern Recognition (CVPR). Minneapolis, Minnesota, USA, June 18–23*, pages 1–8, 2007.

[117] Hristina Hristova, Olivier Le Meur, Rémi Cozot, and Kadi Bouatouch. Style-aware robust color transfer. In *Proceedings of the Workshop on Computational Aesthetics*, CAE '15, pages 67–77, Aire-la-Ville, Switzerland, Switzerland, 2015. Eurographics Association.

[118] Chih-Chung Hsu, Chia-Wen Lin, Yuming Fang, and Weisi Lin. Objective quality assessment for image retargeting based on perceptual distortion and information loss. In *Visual Communications and Image Processing (VCIP), 2013*, pages 1–6, Nov 2013.

[119] R. W. G. Hunt. *The Reproduction of Colour, 5th Edition*. Fountain Press Ltd., 1995.

[120] Yongqing Huo, Fan Yang, Le Dong, and Vincent Brost. Physiological inverse tone mapping based on retina response. *The Visual Computer*, 30(5):507–517, May 2013.

[121] ICC. International Color Consortium, 1998. http://www.color.org/index. xalter.

[122] Fodor Imola, K. A survey of dimension reduction techniques UCRL-ID-148494. Technical report, US Department of Energy, 2002.

[123] Adobe Systems Inc. Photoshop CC, 2013. http://www.adobe.com.

[124] Industrial Light & Magic. OpenEXR, April 2015. http://www.openexr.org/.

[125] P. Irawan, J. A. Ferwerda, and S. R. Marschner. Perceptually based tone mapping of high dynamic range image streams. In *Eurographics Symposium on Rendering, Konstanz, Germany, June 29 – 1 July*, pages 231–242, 2005.

[126] J. Itten. *The Art of Color: The Subjective Experience and Objective Rationale of Color*. A VNR book. Wiley, 1974.

[127] Laurent Itti, Christof Koch, and Ernst Niebur. A model of saliency-based visual attention for rapid scene analysis. *IEEE Trans. Pattern Anal. Mach. Intell.*, 20(11):1254–1259, 1998.

[128] ITU. Recommendation ITU-R BT.709-3: Parameter values for the HDTV standards for production and international programme exchange. International Telecommunications Union, 1998.

[129] ITU. Recommendation ITU-R BT.601-7: Studio encoding parameters of digital television for standard 4:3 and wide-screen 16:9 aspect ratios. International Telecommunications Union, 2011.

[130] ITU. Recommendation ITU-R BT.2020: Parameter values for ultra-high definition television systems for production and international programme exchange. International Telecommunications Union, 2012.

[131] Mica K. Johnson, Kevin Dale, Shai Avidan, Hanspeter Pfister, William T. Freeman, and Wojciech Matusik. Cg2real: Improving the realism of computer generated images using a large collection of photographs. *IEEE Transactions on Visualization and Computer Graphics*, 17(9):1273–1285, July 2011.

[132] Neel Joshi, Wojciech Matusik, Edward H. Adelson, and David J. Kriegman. Personal photo enhancement using example images. *ACM Trans. Graph.*, 29(2):12:1–12:15, April 2010.

[133] D.B. Judd and G. Wyszecki. *Color in Business, Science, and Industry*. Pure and Applied Optics Series. Wiley, 1975.

[134] Tilke Judd, Krista Ehinger, Frédo Durand, and Antonio Torralba. Learning to predict where humans look. In *IEEE International Conference on Computer Vision (ICCV), Kyoto, Japan, September 29 - October 2*, 2009.

[135] Sefy Kagarlitsky, Yael Moses, and Yacov Hel-Or. Piecewise-consistent color mappings of images acquired under various conditions. In *IEEE 12th International Conference on Computer Vision (ICCV), Kyoto, Japan, September 27 – October 4*, pages 2311–2318, 2009.

[136] Sing Bing Kang, Matthew Uyttendaele, Simon Winder, and Richard Szeliski. High dynamic range video. *ACM Trans. Graph.*, 22(3):319–325, July 2003.

[137] Levent Karacan, Erkut Erdem, and Aykut Erdem. Structure-preserving image smoothing via region covariances. *ACM Trans. Graph.*, 32(6):176:1–176:11, November 2013.

[138] Zachi Karni, D. Freedman, and Craig Gotsman. Energy-based image deformation. *Computer Graphics Forum (Proc. of SGP)*, 28(5):1257–1268, July 2009.

[139] Peter Kaufmann, Oliver Wang, Alexander Sorkine-Hornung, Olga Sorkine-Hornung, Aljoscha Smolic, and Markus Gross. Finite element image warping. *Computer Graphics Forum (Proceedings of EUROGRAPHICS)*, 32(2):31–39, 2013.

[140] Min H. Kim and Jan Kautz. Characterization for high dynamic range imaging. *Comptuer Graphics Forum (Proc. of Eurographics)*, 27(2):691–697, April 2008.

[141] Min H. Kim, Tim Weyrich, and Jan Kautz. Modeling human color perception under extended luminance levels. *ACM Trans. Graph.*, 28(3):27:1–27:9, July 2009.

[142] Adam G. Kirk and James F. O'Brien. Perceptually based tone mapping for low-light conditions. *ACM Trans. Graph.*, 30(4):42:1–42:10, July 2011.

[143] Johannes Kopf, Michael F. Cohen, Dani Lischinski, and Matt Uyttendaele. Joint bilateral upsampling. In *ACM SIGGRAPH 2007 Papers*, SIGGRAPH '07, New York, NY, USA, 2007. ACM.

[144] Pavel Korshunov, Philippe Hanhart, Thomas Richter, Alessandro Artusi, Rafal Mantiuk, and Touradj Ebrahimi. Subjective quality assessment database of HDR images compressed with JPEG XT. In *7th International Workshop on Quality of Multimedia Experience (QoMEX), Costa Navarino, Messinia, Greece, May 26–29*, 2015.

[145] Hiroaki Kotera. A scene-referred color transfer for pleasant imaging on display. In *Proceedings of the 2005 International Conference on Image Processing (ICIP). Genoa, Italy, September 11–14*.

[146] Rafael Pacheco Kovaleski and Manuel M. Oliveira. High-quality brightness enhancement functions for real-time reverse tone mapping. *Vis. Comput.*, 25(5–7):539–547, 2009.

[147] Alper Koz and Frederic Dufaux. Methods for improving the tone mapping for backward compatible high dynamic range image and video coding. *Image Commun.*, 29(2):274–292, February 2014.

[148] Philipp Krähenbühl, Manuel Lang, Alexander Hornung, and Markus Gross. A system for retargeting of streaming video. *ACM Trans. Graph.*, 28(5), 2009.

[149] Philipp Krähenbühl, Manuel Lang, Alexander Hornung, and Markus Gross. A system for retargeting of streaming video. *ACM Trans. Graph.*, 28(5):126:1–126:10, December 2009.

[150] Grzegorz Krawczyk, Karol Myszkowski, and Hans-Peter Seidel. Lightness perception in tone reproduction for high dynamic range images. *Computer Graphics Forum (Proceedings of Eurographics 2005)*, 24(3):635–645, 2005.

[151] Joel Kronander, Stefan Gustavson, Gerhard Bonnet, Anders Ynnerman, and Jonas Unger. A unified framework for multi-sensor HDR video reconstruction. *Signal Processing: Image Communications*, 29(2):203–215, 2014.

[152] Jiangtao Kuang, Garrett M. Johnson, and Mark D. Fairchild. iCAM06: A refined image appearance model for HDR image rendering. *J. Vis. Comun. Image Represent.*, 18(5):406–414, October 2007.

[153] Gregory Ward Larson. LogLuv encoding for full-gamut, high-dynamic range images. *Journal of Graphics Tools*, 3(1):15–31, 1998.

[154] Gregory Ward Larson, Holly Rushmeier, and Christine Piatko. A visibility matching tone reproduction operator for high dynamic range scenes. *IEEE Transactions on Visualization and Computer Graphics*, 3(4):291–306, 1997.

[155] Olivier Le Meur, Patrick Le Callet, Dominique Barba, and Dominique Thoreau. A coherent computational approach to model bottom-up visual attention. *IEEE Trans. Pattern Anal. Mach. Intell.*, 28(5):802–817, May 2006.

[156] Anat Levin, Dani Lischinski, and Yair Weiss. Colorization using optimization. *ACM Trans. Graph.*, 23(3):689–694, August 2004.

[157] Yuanzhen Li, Lavanya Sharan, and Edward H. Adelson. Compressing and companding high dynamic range images with subband architectures. *ACM Trans. Graph.*, 24(3):836–844, 2005.

[158] Stephen Lin, Jinwei Gu, Shuntaro Yamazaki, and Heung-Yeung Shum. Radiometric calibration from a single image. In *Proceedings of the 2004 IEEE Conference on Computer Vision and Pattern Recognition (CVPR2004), Volume 2*, pages 938–945, Washington, DC, USA, 2004. IEEE Computer Society.

[159] Stephen Lin and Lei Zhang. Determining the radiometric response function from a single grayscale image. In *Proceedings of the 2005 IEEE Computer Society Conference on Computer Vision and Pattern Recognition (CVPR'05), Volume 2*, pages 66–73, Washington, DC, USA, 2005. IEEE Computer Society.

[160] Tony Lindeberg. *Scale-Space Theory in Computer Vision.* Kluwer Academic Publishers, Norwell, MA, USA, 1994.

[161] Dani Lischinski, Zeev Farbman, Matt Uyttendaele, and Richard Szeliski. Interactive local adjustment of tonal values. *ACM Trans. Graph.*, 25(3):646–653, July 2006.

[162] Anmin Liu, Weisi Lin, Chen Hai, and Chen-Xiong Zhang. Image retargeting quality assessment, May 2015. US Patent 9,025,910.

[163] Ce Liu, Jenny Yuen, and Antonio Torralba. Sift flow: Dense correspondence across scenes and its applications. *IEEE Trans. Pattern Anal. Mach. Intell.*, 33(5):978–994, May 2011.

[164] Tie Liu, Zejian Yuan, Jian Sun, Jingdong Wang, Nanning Zheng, Xiaoou Tang, and Heung-Yeung Shum. Learning to detect a salient object. *Pattern Analysis and Machine Intelligence, IEEE Transactions on*, 33(2):353–367, Feb 2011.

[165] Yiming Liu, Michael F. Cohen, Matthew Uyttendaele, and Szymon Rusinkiewicz. Autostyle: Automatic style transfer from image collections to users' images. *Comput. Graph. Forum*, 4(33):21–31, 2014.

[166] David G. Lowe. Distinctive image features from scale-invariant keypoints. *Int. J. Comput. Vision*, 60(2):91–110, 2004.

[167] Qing Luan, Fang Wen, and Ying-Qing Xu. Color transfer brush. In *Proceedings of the 15th Pacific Conference on Computer Graphics and Applications*, PG '07, pages 465–468, Washington, DC, USA, 2007. IEEE Computer Society.

[168] Eva Lübbe. *Colours in the Mind-Colour Systems in Reality: A formula for Colour Saturation*. Books on Demand GmbH, 2008.

[169] Lindsay W. MacDonald. Gamut mapping in perceptual colour space. In *Color and Imaging Conference Final Program and Proceedings*, volume 4, pages 193–196, January 1993.

[170] Zicong Mai, Hassan Mansour, Rafal Mantiuk, Panos Nasiopoulos, Rabab Kreidieh Ward, and Wolfgang Heidrich. Optimizing a tone curve for backward-compatible high dynamic range image and video compression. *IEEE Transactions on Image Processing*, 20(6):1558–1571, 2011.

[171] Zicong Mai, Hassan Mansour, Panos Nasiopoulos, and Rabab Ward. Visually-favorable tone-mapping with high compression performance. In *17th IEEE International Conference on Image Processing (ICIP)*, pages 1285–1288, 2010.

[172] S. Mann and R. W. Picard. Being undigital with digital cameras: Extending dynamic range by combining differently exposed pictures. In *In Proceedings of IS&T 46th Annual Conference*, pages 422–428, May 1995.

[173] R. Mantiuk, S. Daly, K. Myszkowski, and H-P. Seidel. Predicting visible differences in high dynamic range images: Model and its calibration. In *Proc. of SPIE: Human Vision and Electronic Imaging X*, volume 5666, pages 204–214, 2005.

[174] Radoslaw Mantiuk, Rafał Mantiuk, Anna Tomaszewska, and Wolfgang Heidrich. Color correction for tone mapping. *Computer Graphics Forum (Proc. of EUROGRAPHICS 2009)*, 28(2):193–202, 2009.

[175] RafałMantiuk, Scott Daly, and Louis Kerofsky. Display adaptive tone mapping. *ACM Trans. Graph.*, 27(3):68:1–68:10, August 2008.

[176] Rafał Mantiuk, Alexander Efremov, Karol Myszkowski, and Hans-Peter Seidel. Backward compatible high dynamic range mpeg video compression. *ACM Trans. Graph.*, 25(3):713–723, 2006.

[177] Rafał Mantiuk, Grzegorz Krawczyk, Karol Myszkowski, and Hans-Peter Seidel. Perception-motivated high dynamic range video encoding. *AĊM Trans. Graph.*, 23(3):733–741, 2004.

[178] Rafal Mantiuk, Karol Myszkowski, and Hans-Peter Seidel. A perceptual framework for contrast processing of high dynamic range images. *ACM Trans. Appl. Percept.*, 3(3):286–308, July 2006.

[179] Rafał Mantiuk and Hans-Peter Seidel. Modeling a generic tone-mapping operator. *Computer Graphics Forum*, 27(2):699–708, April 2008.

[180] Rafal Mantiuk, Anna Tomaszewska, and Radoslaw Mantiuk. Comparison of four subjective methods for image quality assessment. *Comput. Graph. Forum*, 31(8):2478–2491, December 2012.

[181] Belen Masia, Sandra Agustin, Roland W. Fleming, Olga Sorkine, and Diego Gutierrez. Evaluation of reverse tone mapping through varying exposure conditions. *ACM Trans. Graph.*, 28(5):1–8, 2009.

[182] Alla Maslennikova and Vladimir Vezhnevets. Interactive local color transfer between images. In *GraphiCon*, 2007.

[183] Y. Matsuda. *Color Design.* Asakura Shoten (Japan), 1995.

[184] Marc Mehl. Picturenaut 3.2, 2012. http://www.hdrlabs.com/picturenaut/.

[185] Tom Mertens, Jan Kautz, and Frank Van Reeth. Exposure fusion. In *PG '07: Proceedings of the 15th Pacific Conference on Computer Graphics and Applications*, pages 382–390, Washington, DC, USA, 2007. IEEE Computer Society.

[186] Laurence Meylan and Sabine Süsstrunk. High dynamic range image rendering with a Retinex-based adaptive filter. *IEEE Transactions on Image Processing*, 15(9):2820–2830, 2006.

[187] Tomoo Mitsunaga and Shree K. Nayar. Radiometric self-calibration. *IEEE Computer Society Conference on Computer Vision and Pattern Recognition (CVPR). Ft. Collins, CO, USA, June 23–25*, 1:1374, 1999.

[188] P. Moon and D.E. Spencer. Visual data applied to lighting design. *J. Opt. Soc. Am.*, 34(605), 1944.

[189] Nathan Moroney, Mark D. Fairchild, Robert W.G. Hunt, Changjun Li, M. Ronnier Luo, and Todd Newman. The CIECAM02 color appearance mode. In *Tenth Color and Imaging Conference.* Society for Imaging Science and Technology, 2002.

[190] Jan Morovic. *Color Gamut Mapping.* John Wiley & Sons Ltd., 2008.

[191] Jan Morovič. *To Develop a Universal Gamut Mapping Algorithm.* PhD Thesis, 1998.

[192] Przemyslaw Musialski, Ming Cui, Jieping Ye, Anshuman Razdan, and Peter Wonka. A framework for interactive image color editing. *Vis. Comput.*, 29(11):1173–1186, November 2013.

[193] R. Ward N. Sun, H. Mansour. HDR image construction from multi-exposed stereo LDR images. In *17th IEEE International Conference on Image Processing (ICIP)*, pages 2973–2976, September 2010.

[194] Manish Narwaria, Matthieu Perreira Da Silva, Patrick Le Callet, and Romuald Ppion. Tone mapping-based high-dynamic-range image compression: study of optimization criterion and perceptual quality. *Optical Engineering*, 52(10):1–15, Oct 2013.

[195] Shree K. Nayar and Tomoo Mitsunaga. High dynamic range imaging: Spatially varying pixel exposures. In *IEEE Conference on Computer Vision and Pattern Recognition (CVPR)*, volume 1, pages 472–479, Jun 2000.

[196] Shree K. Nayar and Srinivasa G. Narasimhan. Assorted pixels: Multi-sampled imaging with structural models. In *ECCV*, July 2002.

[197] Yoshinobu Nayatani. Simple estimation methods for the Helmholtz–Kohlrausch effect. *Color Research and Application*, 22(6):385–401, December 1997.

[198] Yoshinobu Nayatani. Relationship between the two kinds of representation methods in the Helmholtz–Kohlrausch effect. *Journal of the Illuminating Engineering Institute of Japan*, 82(2):143–152, February 1998.

[199] L. Neumann, M. Čadík, and A. Nemcsics. An efficient perception-based adaptive color to gray transformation. In *Proceedings of the Third Eurographics Conference on Computational Aesthetics in Graphics, Visualization and Imaging*, Computational Aesthetics'07, pages 73–80, Aire-la-Ville, Switzerland, Switzerland, 2007. Eurographics Association.

[200] László Neumann and Attila Neumann. Color style transfer techniques using hue, lightness and saturation histogram matching. In *Proceedings of the First Eurographics Conference on Computational Aesthetics in Graphics, Visualization and Imaging*, Computational Aesthetics'05, pages 111–122, Aire-la-Ville, Switzerland, Switzerland, 2005. Eurographics Association.

[201] R. M. H. Nguyen, S. J. Kim, and M. S. Brown. Illuminant aware gamut-based color transfer. *Comput. Graph. Forum*, 33(7):319–328, October 2014.

[202] Peter O'Donovan, Aseem Agarwala, and Aaron Hertzmann. Color compatibility from large datasets. *ACM Trans. Graph.*, 30(4):63:1–63:12, July 2011.

[203] Masahiro Okuda and Nicola Adami. Two-layer coding algorithm for high dynamic range images based on luminance compensation. *J. Vis. Comun. Image Represent.*, 18(5):377–386, 2007.

[204] Aude Oliva and Antonio Torralba. Modeling the shape of the scene: A holistic representation of the spatial envelope. *Int. J. Comput. Vision*, 42(3):145–175, May 2001.

[205] Miguel Oliveira, Angel Domingo Sappa, and Vítor Santos. Unsupervised local color correction for coarsely registered images. In *The 24th IEEE Conference on Computer Vision and Pattern Recognition (CVPR). Colorado Springs, CO, USA, 20-25 June*, pages 201–208, 2011.

[206] Thomas Oskam, Alexander Hornung, Robert W. Sumner, and Markus Gross. Fast and stable color balancing for images and augmented reality. In *Proceedings of the 2012 Second International Conference on 3D Imaging, Modeling, Processing, Visualization & Transmission*, 3DIMPVT '12, pages 49–56, Washington, DC, USA, 2012. IEEE Computer Society.

[207] Ahmet Oğuz Akyüz and Asli Genctav. A reality check for radiometric camera response recovery algorithms. *Computers & Graphics*, 37(7):935–943, 2013.

[208] Daniele Panozzo, Ofir Weber, and Olga Sorkine. Robust image retargeting via axis-aligned deformation. *Computer Graphics Forum (Proceedings of EUROGRAPHICS)*, 31(2):229–236, 2012.

[209] Nicolas Papadakis, Edoardo Provenzi, and Vicent Caselles. A variational model for histogram transfer of color images. *IEEE Transactions on Image Processing*, 20(6):1682–1695, 2011.

[210] Sylvain Paris. Edge-preserving smoothing and mean-shift segmentation of video streams. In *Proceedings of the 10th European Conference on Computer Vision: Part II*, ECCV '08, pages 460–473, Berlin, Heidelberg, 2008. Springer-Verlag.

[211] Sylvain Paris, Samuel W. Hasinoff, and Jan Kautz. Local laplacian filters: Edge-aware image processing with a laplacian pyramid. *ACM Trans. Graph.*, 30(4):68:1–68:12, July 2011.

[212] Sumanta N. Pattanaik, James A. Ferwerda, Mark D. Fairchild, and Donald P. Greenberg. A multiscale model of adaptation and spatial vision for realistic image display. In *Proceedings of the 25th Annual Conference on Computer Graphics and Interactive Techniques*, SIGGRAPH '98, pages 287–298, New York, NY, USA, 1998. ACM.

[213] Sumanta N. Pattanaik, Jack Tumblin, Hector Yee, and Donald P. Greenberg. Time-dependent visual adaptation for fast realistic image display. In *Proceedings of the 27th Annual Conference on Computer Graphics and Interactive Techniques*, SIGGRAPH '00, pages 47–54, New York, NY, USA, 2000. ACM Press/Addison-Wesley Publishing Co.

[214] Federico Perazzi, Philipp Krähenbühl, Yael Pritch, and Alexander Hornung. Saliency filters: Contrast based filtering for salient region detection. In *IEEE Conference on Computer Vision and Pattern Recognition (CVPR). Providence, RI, USA, June 16–21*, pages 733–740, 2012.

[215] Federico Perazzi, Olga Sorkine-Hornung, and Alexander Sorkine-Hornung. Efficient salient foreground detection for images and video using fiedler vectors. In W. Bares, M. Christie, and R. Ronfard, editors, *Eurographics Workshop on Intelligent Cinematography and Editing*. The Eurographics Association, 2015.

[216] Patrick Pérez, Michel Gangnet, and Andrew Blake. Poisson image editing. *ACM Trans. Graph.*, 22(3):313–318, July 2003.

[217] P. Perona and J. Malik. Scale-space and edge detection using anisotropic diffusion. *IEEE Trans. Pattern Anal. Mach. Intell.*, 12(7):629–639, July 1990.

[218] F. Petié and A.C. Kakaram. N-dimensional probability density function transfer and its application to color transfer. In *Tenth IEEE International Conference on Computer Vision (ICCV)*, pages 1434–1439. IEEE, October 2005.

[219] F. Petié, A.C. Kakaram, and R. Dahyot. Automated colour grading using colour distribution transfer. *Computer Vision and Image Understanding*, 107(1-2):123–137, July–August 2007.

[220] Georg Petschnigg, Richard Szeliski, Maneesh Agrawala, Michael Cohen, Hugues Hoppe, and Kentaro Toyama. Digital photography with flash and no-flash image pairs. *ACM Trans. Graph.*, 23(3):664–672, 2004.

[221] F. Pitiánd A. Kokaram. The linear Monge–Kantorovitch colour mapping for example-based colour transfer. In *IEE European Conference on Visual Media Production (CVMP'07)*, London, UK, December 2007.

[222] Tania Pouli, Alessandro Artusi, Francesco Banterle, Ahmet Oğuz Akyüz, Hans-Peter Seidel, and Erik Reinhard. Color correction for tone reproduction. In *Color and Imaging Conference*. Society for Imaging Science and Technology, November 2013.

[223] Tania Pouli and Erik Reinhard. Progressive histogram reshaping for creative color transfer and tone reproduction. In *Proceedings of the 8th International Symposium on Non-Photorealistic Animation and Rendering*, NPAR '10, pages 81–90, New York, NY, USA, 2010. ACM.

[224] Tania Pouli and Erik Reinhard. Extended papers from NPAR 2010: Progressive color transfer for images of arbitrary dynamic range. *Comput. Graph.*, 35(1):67–80, February 2011.

[225] William H. Press, Saul A. Teukolsky, William T. Vetterling, and Brian P. Flannery. *Numerical Recipes 3rd Edition: The Art of Scientific Computing*. Cambridge University Press, New York, NY, USA, 2007.

[226] Julien Rabin, Julie Delon, and Yann Gousseau. Removing artefacts from color and contrast modifications. *IEEE Transactions on Image Processing*, 20(11):3073–3085, 2011.

[227] Juliet Rabin, Julie Delon, and Yann Gousseau. Regularization of transportation maps for color and contrast transfer. In *17th IEEE International Conference on Image Processing (ICIP)*, pages 1933–1936, 2010.

[228] Karl Rasche, Robert Geist, and James Westall. Re-coloring images for gamuts of lower dimension. *Computer Graphics Forum*, pages 423–432, 2005.

[229] Erik Reinhard, Michael Ashikhmin, Bruce Gooch, and Peter Shirley. Color transfer between images. *IEEE Comput. Graph. Appl.*, 21(5):34–41, September 2001.

[230] Erik Reinhard, Michael Stark, Peter Shirley, and James Ferwerda. Photographic tone reproduction for digital images. *ACM Trans. Graph.*, 21(3):267–276, 2002.

[231] Allan G. Rempel, Matthew Trentacoste, Helge Seetzen, H. David Young, Wolfgang Heidrich, Lorne Whitehead, and Greg Ward. Ldr2hdr: On-the-fly reverse tone mapping of legacy video and photographs. *ACM Trans. Graph.*, 26(3):39, 2007.

[232] Thomas Richter, Alessandro Artusi, and Massmiliano Agostinelli. Information technology: Scalable compression and coding of continuous-tone still images, HDR floating-point coding. *ISO/IEC 18477-7*, 2016.

[233] Tobias Ritschel, Matthias Ihrke, Jeppe Revall Frisvad, Joris Coppens, Karol Myszkowski, and Hans-Peter Seidel. Temporal glare: Real-time dynamic simulation of the scattering in the human eye. *Computer Graphics Forum*, 28(2), April 2009.

[234] Mark A. Robertson, Sean Borman, and Robert L. Stevenson. Estimation-theoretic approach to dynamic range enhancement using multiple exposures. *Journal of Electronic Imaging*, 12(2):219–228, April 2003.

[235] Carsten Rother, Vladimir Kolmogorov, and Andrew Blake. GrabCut: Interactive foreground extraction using iterated graph cuts. *ACM Trans. Graph.*, 23(3):309–314, August 2004.

[236] Michael Rubinstein, Diego Gutierrez, Olga Sorkine, and Ariel Shamir. A comparative study of image retargeting. *ACM Transactions on Graphics (Proceedings of ACM SIGGRAPH ASIA)*, 29(5):160:1–160:10, 2010.

[237] Michael Rubinstein, Ariel Shamir, and Shai Avidan. Improved seam carving for video retargeting. *ACM Trans. Graph.*, 27(3), 2008.

[238] Michael Rubinstein, Ariel Shamir, and Shai Avidan. Multi-operator media retargeting. *ACM Trans. Graph.*, 28(3), 2009.

[239] Daniel L. Ruderman, Thomas W. Cronin, and Chuan-Chin Chiao. Statistics of cone responses to natural images: Implications for visual coding. *Journal of the Optical Society of America A*, 15:2036–2045, August 1998.

[240] Anthony Santella, Maneesh Agrawala, Doug DeCarlo, David Salesin, and Michael Cohen. Gaze-based interaction for semi-automatic photo cropping. In *Proceedings of CHI*, pages 771–780, 2006.

[241] Jason M Saragih, Simon Lucey, and Jeffrey Cohn. Face alignment through subspace constrained mean-shifts. In *International Conference of Computer Vision (ICCV)*, pages 1034–1041. IEEE Computer Society, September 2009.

[242] Nikhil Sawant and Niloy J. Mitra. Color harmonization for videos. In *Proceedings of the 2008 Sixth Indian Conference on Computer Vision, Graphics & Image Processing*, ICVGIP '08, pages 576–582, Washington, DC, USA, 2008. IEEE Computer Society.

[243] Christophe Schlick. Quantization techniques for visualization of high dynamic range pictures. In *Proceedings of the Fifth Eurographics Workshop on Rendering*, pages 7–18, June 1994.

[244] Helge Seetzen, Wolfgang Heidrich, Wolfgang Stuerzlinger, Greg Ward, Lorne Whitehead, Matthew Trentacoste, Abhijeet Ghosh, and Andrejs Vorozcovs. High dynamic range display systems. *ACM Trans. Graph.*, 23(3):760–768, August 2004.

[245] C. R. Senanayake and Daniel C. Alexander. Colour transfer by feature based histogram registration. In *Proceedings of the British Machine Vision Conference, University of Warwick, UK, September 10-13*, pages 1–10, 2007.

[246] Ariel Shamir and Olga Sorkine. Visual media retargeting. In *ACM SIGGRAPH ASIA Courses*, 2009.

[247] Feng Shao, Gang-Yi Jiang, Mei Yu, and Yo-Sung Ho. Fast color correction for multi-view video by modeling spatio-temporal variation. *J. Vis. Comun. Image Represent.*, 21(5-6):392–403, July 2010.

[248] Shizhe Shen. *Color Difference Formula and Uniform Color Space Modeling and Evaluation*. PhD thesis, Rochester Institute of Technology, 2009.

[249] YiChang Shih, Sylvain Paris, Connelly Barnes, William T. Freeman, and Frédo Durand. Style transfer for headshot portraits. *ACM Trans. Graph.*, 33(4):148:1–148:14, July 2014.

[250] SIM2. Solar HDR Display. http://www.sim2.com/HDR.

[251] Kaleigh Smith, Pierre-Edouard Landes, Joëlle Thollot, and Karol Myszkowski. Apparent greyscale: A simple and fast conversion to perceptually accurate images and video. *Comput. Graph. Forum*, 27(2):193–200, April 2008.

[252] Stephen M. Smith and J. Michael Brady. SUSAN: A new approach to low level image processing. *Int. J. Comput. Vision*, 23(1):45–78, May 1997.

[253] Olga Sorkine and Marc Alexa. As-rigid-as-possible surface modeling. In *Proceedings of EUROGRAPHICS/ACM SIGGRAPH Symposium on Geometry Processing*, pages 109–116, 2007.

[254] L. Spillmann and J. S. Werner. *Visual Perception The Neurophysiological Foundations.* Elsevier, 1990.

[255] Thomas G. Stockham. Image processing in the context of a visual model. *Proceedings of the IEEE*, 60(7):828–842, July 1972.

[256] Zhuo Su, Xiaonan Luo, and Alessandro Artusi. A novel image decomposition approach and its applications. *The Visual Computer*, 29(10):1011–1023, 2013.

[257] Kartic Subr, Cyril Soler, and Frédo Durand. Edge-preserving multiscale image decomposition based on local extrema. *ACM Trans. Graph.*, 28(5):147:1–147:9, December 2009.

[258] Kalyan Sunkavalli, Micah K. Johnson, Wojciech Matusik, and Hanspeter Pfister. Multi-scale image harmonization. *ACM Trans. Graph.*, 29(4):125:1–125:10, July 2010.

[259] Richard Szeliski. *Computer Vision: Algorithms and Applications.* Springer-Verlag New York, Inc., New York, NY, USA, 1st edition, 2010.

[260] Yu-Wing Tai, Jiaya Jia, and Chi-Keung Tang. Local color transfer via probabilistic segmentation by expectation-maximization. In *Proceedings of the IEEE Computer Society Conference on Computer Vision and Pattern Recognition (CVPR), Volume 1*, pages 747–754, Washington, DC, USA, 2005. IEEE Computer Society.

[261] Yu-Wing Tai, Jiaya Jia, and Chi-Keung Tang. Soft color segmentation and its applications. *IEEE Trans. Pattern Anal. Mach. Intell.*, 29(9):1520–1537, 2007.

[262] Z. Tang, Z. Miao, and Y. Wan. Image composition with color harmonization. In *25th International Conference of Image and Vision Computing New Zealand (IVCNZ)*, pages 1–8. IEEE, 2010.

[263] Z. Tang, Z. Miao, Y. Wan, and F.F Jesse. Colour harmonisation for images and videos via two-level graph cut. *Image Processing, IET*, 5(7):630–643, October 2011.

[264] Sensate Technologies. White paper autobrite imaging technology: Wide dynamic range for automotive machine vision. Technical report, 2007.

[265] Josh B. Tenenbaum, V. De Silva, and John C. Langford. A global geometric framework for nonlinear dimensionality reduction. *Science*, 5500(290):2319–2323, 2000.

[266] Michael D. Tocci, Chris Kiser, Nora Tocci, and Pradeep Sen. A versatile HDR video production system. *ACM Transactions on Graphics*, 30(4):41:1–41:10, July 2011.

[267] M. Tokumaru, N. Muranaka, and S. Imanishi. Color design support system considering color harmony. In *Proceedings of the 2002 IEEE International Conference on Fuzzy Systems (FUZZ-IEEE'02)*, pages 378–383. IEEE, 2002.

[268] Carlo Tomasi and Roberto Manduchi. Bilateral filtering for gray and color images. In *ICCV '98: Proceedings of the Sixth International Conference on Computer Vision*, page 839, Washington, DC, USA, 1998. IEEE Computer Society.

[269] Jack Tumblin, Amit Agrawal, and Ramesh Raskar. Why I want a gradient camera. In *IEEE Conference on Computer Vision and Pattern Recognition (CVPR)*, volume 1, pages 103–110, June 2005.

[270] Jack Tumblin, Jessica K. Hodgins, and Brian K. Guenter. Two methods for display of high contrast images. *ACM Trans. Graph.*, 18(1):56–94, 1999.

[271] Okan Tarhan Tursun, Ahmet Oğuz Akyüz, Aykut Erdem, and Erkut Erdem. The state of the art in HDR deghosting: A survey and evaluation. *Computer Graphics Forum (Proc. of Eurographics)*, 34(2):683–707, May 2015.

[272] Keiji Uchikawa and Robert M. Boynton. Categorical color perception of japanese observers: Comparison with that of Americans. *Vision Research*, 27(10):1824–1833, 1987.

[273] T. Nguyen V. Ramachandra, M. Zwicker. HDR imaging from differently exposed multiview videos. In *3DTV Conference: The True Vision-Capture, Transmission and Display of 3D Video*, May 2008.

[274] Vladimir Vezhnevets and Vadim Konushin. GrowCut: Interactive multi-label n-d image segmentation by cellular automata. In *GraphiCon*, 2005.

[275] Paul Viola and Michael J. Jones. Robust real-time face detection. *Int. J. Comput. Vision*, 57(2):137–154, 2004.

[276] Elena Šikudova, Tania Pouli, Alessandro Artusi, Akyüz Ahmet Oğuz, Banterle Francesco, Erik Reinhard, and Zeynep Miray Mazlumoglu. A gamut mapping framework for color-accurate reproduction of HDR images. *IEEE Transaction of Computer Graphics and Applications*, 2015.

[277] Baoyuan Wang, Yizhou Yu, Tien-Tsin Wong, Chun Chen, and Ying-Qing Xu. Data-driven image color theme enhancement. *ACM Trans. Graph.*, 29(6):146:1–146:10, December 2010.

[278] Baoyuan Wang, Yizhou Yu, and Ying-Qing Xu. Example-based image color and tone style enhancement. In *ACM SIGGRAPH 2011 Papers*, SIGGRAPH '11, pages 64:1–64:12, New York, NY, USA, 2011. ACM.

[279] Hongcheng Wang, Ramesh Raskar, and Narendra Ahuja. High dynamic range video using split aperture camera. In *IEEE 6th Workshop on Omnidirectional Vision, Camera Networks and Non-classical Cameras (OMNIVIS, in conjunction with ICCV*, 2005.

[280] Lvdi Wang, Li-Yi Wei, Kun Zhou, Baining Guo, and Heung-Yeung Shum. High dynamic range image hallucination. In *SIGGRAPH '07: ACM SIGGRAPH 2007 Sketches*, page 72, New York, NY, USA, 2007. ACM.

[281] Qi Wang, Xi Sun, and Zengfu Wang. A robust algorithm for color correction between two stereo images. In *Proceedings of the 9th Asian Conference on Computer Vision - Volume Part II*, ACCV'09, pages 405–416, Berlin, Heidelberg, 2010. Springer-Verlag.

[282] Yu-Shuen Wang, Chiew-Lan Tai, Olga Sorkine, and Tong-Yee Lee. Optimized scale-and-stretch for image resizing. *ACM Trans. Graph.*, 27(5), 2008.

[283] Zhou Wang, A. C. Bovik, H. R. Sheikh, and E. P. Simoncelli. Image quality assessment: From error visibility to structural similarity. *Trans. Img. Proc.*, 13(4):600–612, April 2004.

[284] Greg Ward. Real pixels. *Graphics Gems*, 2:15–31, 1991.

[285] Greg Ward and Maryann Simmons. JPEG-HDR: A backwards-compatible, high dynamic range extension to JPEG. In *CIC 13th: Proceedings of the Thirteenth Color Imaging Conference*. The Society for Imaging Science and Technology, 2005.

[286] Gregory Ward. Real pixels. *Graphics Gems*, 2:15–31, 1991.

[287] A. B. Watson. The Cortex transform: Rapid computation of simulated neural images. *Comp. Vision Graphics and Image Processing*, 39:311–327, 1987.

[288] Tomihisa Welsh, Michael Ashikhmin, and Klaus Mueller. Transferring color to greyscale images. *ACM Trans. Graph.*, 21(3):277–280, July 2002.

[289] Chung-Lin Wen, Chang-His Hsieh, Bing-Yu Chen, and Ming Ouhyoung. Example-based multiple local color transfer by strokes. *Computer Graphics Forum*, 27(7):1765–1772, 2008. Pacific Graphics 2008 Conference Proceedings.

[290] Bennett Wilburn, Neel Joshi, Vaibhav Vaish, Eino-Ville Talvala, Emilio Antunez, Adam Barth, Andrew Adams, Mark Horowitz, and Marc Levoy. High performance imaging using large camera arrays. *ACM Trans. Graph.*, 24(3):765–776, July 2005.

[291] Virginia Bowman Wissler. *Illuminated Pixels: The Why, What, and How of Digital Lighting*. CENGAGE Learning, August 2013.

[292] Lior Wolf, Moshe Guttmann, and Daniel Cohen-Or. Non-homogeneous content-driven video-retargeting. In *IEEE International Conference on Computer Vision (ICCV)*, 2007.

[293] Fuzhang Wu, Weiming Dong, Yan Kong, Xing Mei, Jean-Claude Paul, and Xiaopeng Zhang. Content-based colour transfer. *Comput. Graph. Forum*, 32(1):190–203, 2013.

[294] Huisi Wu, Yu-Shuen Wang, Kun-Chuan Feng, Tien-Tsin Wong, Tong-Yee Lee, and Pheng-Ann Heng. Resizing by symmetry-summarization. *ACM Trans. Graph.*, 29(6):159:1–159:10, December 2010.

[295] G. Wyszecki and W.S. Stiles. *Color Science: Concepts and Methods, Quantitative Data and Formulae*. Wiley Series in Pure and Applied Optics. Wiley, 2000.

[296] Yao Xiang, Beiji Zou, and Hong Li. Selective color transfer with multi-source images. *Pattern Recogn. Lett.*, 30(7):682–689, May 2009.

[297] Rong Xiao, Huaiyi Zhu, He Sun, and Xiaoou Tang. Dynamic cascades for face detection. In *IEEE 11th International Conference on Computer Vision (ICCV)*, pages 1–8, Rio de Janeiro, Brazil, 2007.

[298] Xuezhong Xiao and Lizhuang Ma. Color transfer in correlated color space. In *Proceedings of the 2006 ACM International Conference on Virtual Reality Continuum and Its Applications*, VRCIA '06, pages 305–309, New York, NY, USA, 2006. ACM.

[299] XueZhong Xiao and Lizhuang Ma. Gradient-preserving color transfer. *Computer Graphics Forum*, 28(7):1879–1886, 2009.

[300] Li Xu, Cewu Lu, Yi Xu, and Jiaya Jia. Image smoothing via l0 gradient minimization. *ACM Trans. Graph.*, 30(6):174:1–174:12, December 2011.

[301] Shuchang Xu, Yin Zhang, Sanyuan Zhang, and Xiuzi Ye. Uniform color transfer. In *IEEE International Conference on Image Processing ICIP*, pages 940–943. IEEE, September 2005.

[302] Chuan-Kai Yang and Li-Kai Peng. Automatic mood-transferring between color images. *IEEE Comput. Graph. Appl.*, 28(2):52–61, March 2008.

[303] Jimei Yang and Ming-Hsuan Yang. Top-down visual saliency via joint crf and dictionary learning. In *IEEE Conference on Computer Vision and Pattern Recognition (CVPR). Providence, RI, USA, June 16–21*, pages 2296–2303, June 2012.

[304] Leonid P. Yaroslavsky. *Digital Picture Processing, An Introduction*. Springer-Verlag, 1985.

[305] Hojatollah Yeganeh and Zhou Wang. Objective quality assessment of tone mapped images. *IEEE Transactions on Image Processing*, 22(2):657–667, February 2013.

[306] Guo-Xin Zhang, Ming-Ming Cheng, Shi-Min Hu, and Ralph R. Martin. A shape-preserving approach to image resizing. *Comput. Graph. Forum*, 28(7):1897–1906, 2009.

[307] Yi Zhang, Chun Chen, Jiajun Bu, and Mingli Song. A kernel based algorithm for fast color-to-gray processing. In *Congress on Image and Signal Processing (CISP '08)*, pages 451–455, 2008.

Index

##